Outline of Crystallog
for Biologists

Outline of Crystallography for Biologists

David Blow

Senior Research Fellow
Imperial College of Science,
Technology and Medicine

OXFORD

UNIVERSITY PRESS

Great Clarendon Street, Oxford OX2 6DP

Oxford University Press is a department of the University of Oxford.
It furthers the University's objective of excellence in research, scholarship,
and education by publishing worldwide in

Oxford New York

Auckland Cape Town Dar es Salaam Hong Kong Karachi
Kuala Lumpur Madrid Melbourne Mexico City Nairobi
New Delhi Shanghai Taipei Toronto

With offices in

Argentina Austria Brazil Chile Czech Republic France Greece
Guatemala Hungary Italy Japan South Korea Poland Portugal
Singapore Switzerland Thailand Turkey Ukraine Vietnam

Oxford is a registered trade mark of Oxford University Press
in the UK and in certain other countries

Published in the United States
by Oxford University Press Inc., New York

© Oxford University Press, 2002

British Library Cataloguing in Publication Data

Data available

Library of Congress Cataloging in Publication Data

ISBN 978 0 19 851051 2

10 9 8

Typeset by Newgen Imaging Systems (P) Ltd, Chennai, India
Printed in Great Britain
on acid-free paper by Biddles Ltd, King's Lynn, Norfolk

Contents

Preface vii

Introductory note ix

Part 1 **Fundamentals** 1

  1 **Images and X-rays** 3

  2 **Crystals and symmetry** 22

  3 **Waves** 45

  4 **Diffraction** 57

  5 **Diffraction by crystals** 82

Part 2 **Practice** 103

  6 **Intensity measurements** 105

  7 **Isomorphous replacement** 121

  8 **Anomalous scattering** 147

  9 **Molecular replacement** 162

  10 **Density-modification procedures: improving a calculated map before interpretation** 177

  11 **Electron-density maps** 191

  12 **Structural refinement** 205

  13 **Accuracy of the final model** 222

  References 229

  Index 233

Preface

I dedicate this book to Max Perutz, who built on Desmond Bernal's vision and Dorothy Hodgkin's determination, and created a practical method for visualizing macromolecular structure. He was my guiding (but never directing) PhD supervisor. Later he became the inspirational head of the hugely successful Laboratory of Molecular Biology in Cambridge. He called himself its Chairman, never the Director. It was a privilege to have worked under him and with him.

Fifty years ago macromolecular structure analysis had never been done, and many thought it impossible. Now it is a daily routine, and the book owes most to those whose crucial contributions have brought this about. I make no attempt to separate the key discoveries that represent the many steps that led along the road, and to identify 'who did what'. I believe my colleagues who have pioneered those discoveries will understand why. The book would have been littered with references, it would have been necessary to refer to details and difficulties which now seem insignificant, and some questions of originality and priority might have been hard to resolve. Such details would be inappropriate in this book which is intended to be no more than a simple outline of the subject.

I owe a debt of gratitude to those who have helped with this publication. They have generously allowed me to mention their work and to use their illustrations, often going to great lengths to provide high-quality figures. Many have read chapters, and commented helpfully on my drafts. To illustrate points that are rarely published in detail, I have had to take examples from doctoral theses, and I am particularly grateful to my former students, and others, who have allowed me to use their earliest research work in this way.

Special acknowledgment is due to two professional artists, who have created a few beautiful drawings for me—Neil Powell and Gerard Lindley. Another special notice is due to Ken Holmes, co-author of a review we wrote together in 1965, and which is, in some senses, the parent of this volume, and to his wife Mary who drew the original duck. Rachel Leveridge drew the duck which appears repeatedly in this volume. I must take responsibility for the less artistic figures.

I am grateful to all these helpers, many of them good friends, but including some whom I have never met. A few of them are named below.

Alan Wonacott
Alexy Teplyakov
Alice Vrielink
Alwyn Jones
Amanda Brindley
Andy Leslie
Anne Bloomer
Axel Brünger
Bram Schierbeek (Nonius)
Charles Taylor
Chris Hammond
Christine Muchmore (Mar)
Daniel Schlieper
David Rice
Don Caspar
Edward Hollingsworth
Eleanor Dodson
Elspeth Garman
Emmanuel Saridakis
Frank James
Gerard Lindley

Guy Pallardy
Helen Price
Ian Johnston
Irene Gonsalvez
Jean-René Regnard
Jenny Littlechild
Jim Raftery
John Helliwell
John Squire
Jonathan Goldberg
Katy Brown
Keith Watenpaugh
Kosuke Morikawa
Linda Britton
Lingling Chen
Lyle Jensen
Marie-Claude Moissenet
Max Perutz
Micha Isupov
Michael Rossmanm
Michael Woolfson

Naoki Kinushima
Neil Powell
Nobutoshi Ito
Peter Brick
Peter Main
Rachel Leveridge
Richard Bytheway (Bede
   Scientific)
Richard Wellberry
Silvia Onesti
Simon Phillips
Stephen Bragg
Stephen Lee
Steve Harrison
Steve Lipson
Uli Arndt
V. Ramakrishnan
Wayne Hendrickson
Wolfram Saenger

David Blow
Appledore, Devon
October 2001

Introductory note

This is an elementary non-mathematical text, prolifically illustrated and supplemented by more precise information in 'boxes', which the basic reader can ignore if preferred.

The elementary text is meant to be readable by an undergraduate student with minimal mathematical insight. The only algebraic equations in the main text represent the Bragg equation. The boxes are designed to be brief and clear, but contain some mathematics. There are special sections for students who need help to understand the mathematical expression of a wave, complex numbers and Fourier series in the boxes of Chapter 3.

The first part 'Fundamentals' presents the underlying ideas that allow X-ray structure analysis to be carried out. It would supplement existing biochemistry undergraduate courses, such as the one at Imperial College. The second part 'Practice' gives more information about how the work is done. This might supplement undergraduate project work, or give an introductory background to a postgraduate. The needs of those who may use X-ray diffraction results without an understanding of the techniques and their limitations, has been borne in mind throughout.

The emphasis in this second half is to offer insight into the quality measures of X-ray diffraction analysis, to give the reader some critical insights into the quality and accuracy of a structure determination, to enable him or her to appreciate more clearly which parts of a structure determination may have caused special difficulty.

There is no pretence of completeness. Many matters discussed in standard crystallography textbooks are omitted. The theory is limited to the essentials needed to comprehend the measures of quality. Although practical matters are discussed where relevant, this is in no way a training course in crystallography. Nevertheless, it might be a useful reference book for practising crystallographers!

Part I

Fundamentals

1

Images and X-rays

Max Perutz has written

The X-ray study of proteins is sometimes regarded as an abstruse subject comprehensible only to specialists, but the basic ideas underlying our work are so simple that some physicists find them boring.

(Perutz 1992)

Crystallography provides the most direct way of forming images of molecules. Using crystallography, three-dimensional images have been made of thousands of macromolecules, especially proteins and nucleic acids. These give detailed information about their activity, their mechanism for recognizing and binding substrates and effectors, and the conformational changes which they may undergo. The three-dimensional structures show graphically the evolutionary relationships between molecules from widely separated systems. They give a wide view of the resemblances between different proteins, showing strong links in three-dimensional structure where the relationship between amino-acid sequences has dwindled to insignificance.

The aim of this book is to tell the ordinary biologist enough about the methods of crystallography and the results that it gives, to allow results to be considered critically, to give insight into the limits of interpretation which are possible, and to identify the causes of the limitation in a particular case. Although the basic ideas are simple, the structural results are complicated, and based on huge numbers of measurements. The challenge for the teacher is to find a way through the simple basic ideas, without getting lost in detail.

The main text is purely descriptive, and can be read from start to finish without any detailed mathematical knowledge. The explanations are backed up by numerous illustrations. For the reader who wants more detail, 'boxes' are supplied with more complete information, which can offer a deeper level of understanding, using some clearly presented mathematics.

The book is not intended to teach the reader how to do practical crystallography. There are many other excellent books for this purpose. It is intended, rather, to give an overview which will allow any biochemist or molecular biologist to read structural papers with a critical awareness.

In the first part of the book (Chapters 1–5), the general principles underlying the use of X-rays, crystals and diffraction are presented, using illustrations to make many of the ideas more graphic. In the second part (Chapters 6–13), the steps that need to be followed in the course of structure determination of a crystal are considered in more detail, again with many illustrations. The various measures that are used as the investigation progresses, to assess the quality of the structure determination, are explained. Many of these measures are usually presented as a table in a structure analysis paper (often referred to as 'Table 1'). By following examples, the reader will learn how to use this table to identify parts of the study which may have caused difficulty, and how to judge the accuracy of the results at each stage.

Because of the large volumes of data involved, macromolecular X-ray crystallography has been totally dependent on computers from the start. Computer programs develop very rapidly, and equally they may be overtaken by others overnight. For this reason, individual computer programs are not mentioned here. The aim is to explain what any of them must do, not to give instructions on how to run them.

Magnified images

Protein crystallography is used to produce magnified images of protein molecules. To begin with, let us explore what can be done with ordinary magnifiers, and consider what needs to be done to observe such small objects as atoms and protein molecules.

The unaided eye has difficulty in distinguishing details less than about 0.1 mm (100 μm) in size. A lens enhances the image, and using a carefully mounted lens, Antonie van Leeuwenhoek was able to observe living cells down to about 2 μm in the 1670s. This is already not far off the limit of performance. The best optical microscopes fail to separate details less than about 0.5 μm (500 nm) apart.

Visible light is composed of electromagnetic waves. Our eyes respond to colours over a range from about 350 nm (far violet) to 700 nm (deep red). The details that can be separated in the best microscopes are about one wavelength apart. Even an ideal microscope could not resolve details less than half a wavelength apart. The wavelength defines the fineness of detail that can be observed.

If an image is to be made on the atomic scale, a 'light' (electromagnetic wave) of an appropriate wavelength must be used. Atoms that are covalently bonded to each other are 1–2 Å apart; strong polar interactions and hydrogen bonds occur when atoms are 2.5–3.5 Å apart. In order to image the atomic structure of a molecule, it is necessary to use wavelengths no larger than a couple of Ångström units. Electromagnetic waves of this wavelength are known as X-rays.

(Note that one Ångström unit (Å) is 0.1 nm or 10^{-10} m. It is not a primary SI unit, but according to the SI it is an acceptable alternative measure to express a length. It is convenient because it is the order of magnitude of an interatomic bond length, and it is still used most often in descriptions of molecular structure. For these reasons the Ångström is used here, but everyone should be able to convert into nanometres without hesitation: 10 Å make 1 nm.)

X-rays

Electromagnetic waves are disturbances of the electric and magnetic fields. In a pure monochromatic light wave, the electric and magnetic fields vary regularly across the direction in which they are travelling. A 'pure' waveform has a characteristic shape, illustrated in Fig. 1.1, which explains the meaning of *wavelength* and *amplitude* of the waveform. The shape of this waveform is called *sinusoidal* because its shape is exactly that of a graph of sine x or cosine x. In an electromagnetic wave, the electric field and the magnetic field both vary sinusoidally.

Electromagnetic waves span a huge spectrum, from long-wave radio waves, the wavelength of which may be several kilometres, down to γ-rays, the wavelength of which is much smaller than the electronic orbitals of atoms. In many parts of this spectrum the waves are strongly absorbed by matter, but Fig. 1.2 shows the three convenient 'windows' for observation and communication.

The radio/television/microwave range is the window at long wavelength. These waves travel freely through air, and some other kinds of matter, but are absorbed by water and other polarizable materials. In this range, waves may be made and detected by electronic equipment which shapes the waves, or follows their shapes, in full detail, making systems ideal for communication.

There is a range in the near infrared/visible/near ultraviolet wavelengths, where both air and water are transparent. This wavelength range is exploited by many kinds of organism for vision, including ourselves. It is used for the conventional forms of spectroscopy. The waves are vibrating too quickly for an electronic detector to follow every detail of the waveform. Human colour vision is achieved by sensing the quantity of radiation absorbed in three different kinds of pigment, whose absorption spectra peak at different wavelengths.

At wavelengths less than 250 nm (far ultraviolet), water becomes strongly absorbing, and this prevents propagation of the waves in biological tissues. At slightly shorter

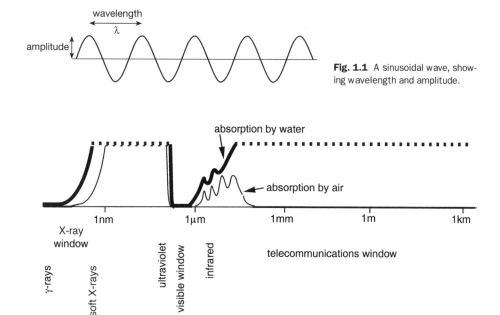

Fig. 1.1 A sinusoidal wave, showing wavelength and amplitude.

Fig. 1.2 Useful windows in the electromagnetic spectrum (schematic).

wavelengths, air also absorbs strongly. As the wavelength is further reduced, the radiation becomes more penetrating: between 10 Å and 1 Å air, and then water, become transparent (and at shorter wavelengths, other materials are also transparent). These penetrating *X-rays* were discovered by Wilhelm von Röntgen because they exposed photographic film encased in a lightproof container.

X-rays having a wavelength of more than about 2 Å are known as 'soft' X-rays. They are significantly absorbed by air and strongly by water. If the wavelength is ten times shorter, say 0.2 Å or less, X-rays are very penetrating. They interact so weakly with matter that a rather large sample is needed to scatter an observable amount of energy. As Röntgen discovered, they penetrate biological tissues and are convenient for radiography of the human body (Fig. 1.3). X-rays of even shorter wavelength are emitted in a variety of nuclear disintegrations, and are known as γ-rays.

The range of wavelengths between about 0.5 Å and 1.6 Å is the most suitable for X-ray crystallography. These waves are sufficiently penetrating to study samples up to a millimetre or so in size, but are scattered strongly by matter. Their wavelength is just right for seeing atoms.

Fig. 1.3 Radioscopic examination of a patient in 1896. (Painting by Jacques Rohr MD (1985), reproduced from Pallardy *et al.* (1989) by permission of the authors.)

X-ray microscopes

Great effort has gone into making devices like a microscope, which can collect the X-rays scattered from an object and use a lens or some comparable device to form an image. There are several reasons why lenses are not available to study structure in atomic detail. The most obvious is the impossibility of forming a lens or mirror from any material composed of atoms, which could be accurately shaped to form a good image at wavelengths of atomic size. X-ray lenses are also impracticable because, for X-rays, the refractive index of all materials is very close to that of a vacuum. Using zone plates, X-ray microscopes have been devised for wavelengths down to 50 Å or so, but these cannot study detailed molecular structure.

Another fundamental problem is the chemical damage done to matter by X-rays, which would cause a molecule to be destroyed before it could be imaged at atomic resolution. By working with crystals, we can look at the scattering from a huge array of molecules, and destruction of a few of them is not so serious. We shall return to this later. Similar, but less severe, problems are encountered in electron microscopy.

A lens is a device that collects all the scattered rays that fall on it, and turns them to bring them together again (Fig. 1.4). In this way, all the scattered rays overlap at the image. The way they overlap depends on precisely where the peaks of the different waves are at any instant, and it is this relationship of the peaks of the waves which causes the image to resemble the scattering object.

If we could measure the properties of the scattered rays precisely, we might use a computer, instead of a lens, to figure out how the scattered rays would overlap to make the image. But in reality, we can only measure the energy scattered in each direction. This energy corresponds to knowing the amplitude of the scattered wave. But there is no way to observe the time of arrival of the peaks of X-rays scattered in different directions, and this prevents us from calculating what the image looks like.

This property, related to the time of arrival of the peak of a wave, is called its *phase*. Two waves of the same wavelength, with different phase, are illustrated in Fig. 1.5. Because no instruments exist which directly observe the phases of X-rays, crystallographers have to use indirect methods to find the information. More information about phases will be given in Chapter 3, and methods of solving the *phase problem* are addressed in Chapters 7, 8, and 9.

Since a conventional microscope using a lens cannot make the image we need, structural research is forced to use other methods. It has been found that the scattering of X-rays by *crystals* can be manipulated to give the necessary

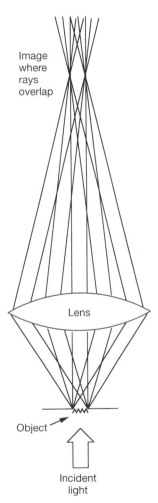

Image where rays overlap

Lens

Object

Incident light

Fig. 1.4 Light scattered from the object is intercepted by the lens. A magnified image of the object is produced where the scattered rays are brought together.

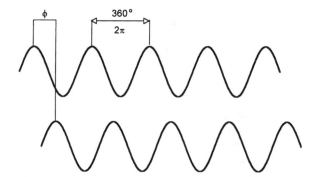

Fig. 1.5 Two waves whose phase differs by an angle ϕ. A whole wavelength corresponds to a phase change of 360° or 2π radians.

information. Crystals provide an ordered array in which every molecule has the same scattering properties. In Chapter 2 we come back to the important features of crystals.

X-ray images

Before Max von Laue discovered X-ray diffraction in 1912, chemists already had clear ideas about the shapes of molecules of organic compounds. Van't Hoff and LeBel had suggested independently in 1862 that the four valences of carbon point in equally spaced directions, corresponding to the shape of a regular tetrahedron (Fig 1.6). Kekulé proposed in 1865 that the carbons of benzene were linked in a special way, so that the six equivalent carbon atoms formed a flat ring (Fig. 1.7). Emil Fischer used the four carbon valencies to propose the nature of optical isomers, and to suggest shapes for sugar molecules (Fig. 1.8). But all these ideas were mental pictures not based on experimental evidence.

In 1912, Max von Laue's laboratory assistants Walter Friedrich and Paul Knipping followed his instructions to place a zinc blende crystal in a narrow beam of X-rays, and to place a photographic film a few centimetres behind it (Fig. 1.9). They quickly improved the experimental set-up to show clearly that the crystal was diffracting (or scattering) the X-ray beam into specific directions (Fig. 1.10). Within a year of von Laue's discovery, Lawrence Bragg interpreted the X-ray diffraction from crystals of zinc sulphide, sodium chloride and potassium chloride, to give clear evidence about the three-dimensional arrangement of atoms (Fig. 1.11). The basis of the methods is outlined in Chapter 5. The structures were so simple (and Bragg's insight so profound) that he could create the image by reasoning. Bragg went directly from observing the crystal X-ray diffraction to models for these structures, the fundamental correctness of which is unquestioned. Structural analysis was extended to more complex structures, but was delayed by the huge computational effort needed to create an image. But by 1923, Dickinson and Raymond were able to exploit a symmetrical molecule, hexamethylene tetramine, to show a molecular shape which confirmed the tetrahedral arrangement of bonds in an organic molecule (Fig. 1.12).

Now ample computational facilities exist, and the development of computer programs for 50 years has provided many tools to take the place of a lens and create an image of a

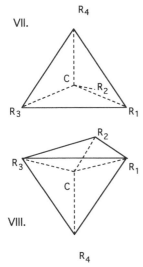

Fig. 1.6 Structural diagrams VII and VIII redrawn from van't Hoff's 1874 pamphlet.

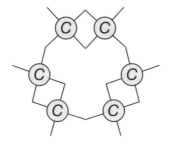

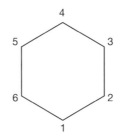

Fig. 1.7 Kekulé's ideas about benzene, from his *Lehrbuch der organischen Chemie* (Kekulé 1866).

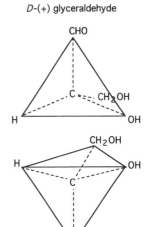

Fig. 1.8 Fischer's chiral convention applied to glyceraldehyde.

Fig. 1.9 Friedrich and Knipping's experiment (courtesy of Deutches Museum, Munich).

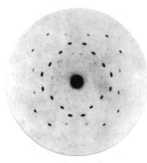

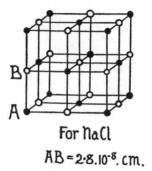

For naCl

$AB = 2.8 . 10^{-8} . cm.$

Fig. 1.10 Diffraction photograph of zinc blende obtained by von Laue and his assistants in 1912. (Reproduced by permission of the Deutsches Museum, Munich, from Hammond (1997).)

Fig. 1.11 Lawrence Bragg's depiction of the structure of sodium chloride. (Reproduced by permission of the Royal Society from Bragg (1913).)

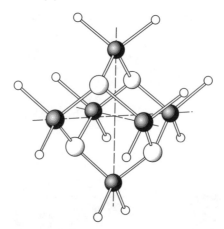

Fig. 1.12 Hexamethylene tetramine as illustrated by Dickinson and Raymond (1923). (Reprinted with permission from Dickinson and Raymond (1923). Copyright 1923, American Chemical Society.)

complex molecule from the X-rays scattered from crystalline samples. Once the phase problem is solved, even approximately, modelling of the structure can begin, and the results can be refined (Chapters 10, 11, and 12).

The properties of X-rays

Electromagnetic waves exist in packets known as photons. It may appear confusing to talk of photons, and then refer to the frequency and wavelength associated with them. Are these things waves or particles? This question troubled Isaac Newton, who wrote about corpuscles of light, and has puzzled many after him. It was a matter of active discussion between Lawrence Bragg and his father William Bragg at the time that Max von Laue's laboratory produced the first X-ray diffraction photographs in 1912.

In those years the first version of the quantum theory was still being developed. Since then, it has become accepted that the photon has particle-like aspects and wave-like

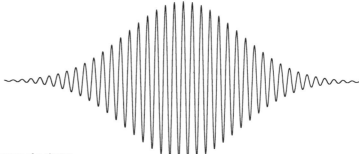

Fig. 1.13 Schematic diagram of a photon.

Box 1.1 Photon energy and wavelength

Planck's formula expresses the energy of a particle or photon E in terms of the frequency, ν, of its wave vibration

$$E = h\nu.$$

The quantity h is known as Planck's constant, and is a fundamental quantity in quantum theory.

For all waves, the velocity of propagation, c, is linked to the wavelength, λ, and the frequency by:

$$c = \nu\lambda.$$

For electromagnetic waves, c is the velocity of light, another fundamental physical constant. Combining these two equations gives:

$$E = hc/\lambda.$$

This formula gives the energy of a photon of given wavelength. The result may be expressed as a potential difference (electron volts) by dividing by the charge e on an electron:

$$V = E/e = hc/e\lambda.$$

This expression gives the voltage difference through which an electron needs to be accelerated to give it the same energy as a photon of given wavelength. Such voltages are convenient expressions of energy on the atomic level, and are used, for example, to express the energy needed to ionize an atom (the ionization potential).

After inserting the values of the constants, if λ is measured in Å and V in volts,

$$V[\text{volts}] = 12\,398/\lambda\,[\text{Å}].$$

To give an electron just enough energy to excite a 1 Å X-ray photon, it needs to be accelerated through 12.4 kV, which is several hundred times the ionization potentials of noble gases. An X-ray photon can produce a trail of energetic ionized atoms when absorbed in matter.

aspects. It may be helpful to think of a photon as a little bunch of waves, travelling at the velocity of light (Fig. 1.13). The particle aspect of X-rays is stressed in this chapter, but elsewhere the wave aspect may appear more important.

The energy carried in one photon is related to its wavelength by Planck's formula (Box 1.1). Photons in the 0.5 Å to 2.0 Å range have energies ranging from 25 000 to 6000 eV: equivalent to that of an electron which has been accelerated through several thousand volts. These energies are more than sufficient to ionize atoms, and for this reason X-rays are classified as 'ionizing radiations'.

There is a matter of great practical importance here. When an X-ray photon is absorbed by an atom, the energy absorbed is more than enough to ionize it, and often to bring it to a free radical form, a much more excited state than that of an ion, which can stimulate a variety of unusual chemical reactions. **X-ray beams rapidly produce blisters and burns on animal tissue. X-rays also act as mutagens and carcinogens. They are dangerous. X-ray exposure to the gonads should be particularly avoided.** Figure 1.3 is disturbing to us today, because both the patient and her radiologist are being exposed to totally unacceptable doses of radiation.

X-ray diffraction apparatus is therefore maintained in an enclosure, made of material that the X-rays cannot penetrate, and interlocked so that the X-ray beam must be closed off before a person can enter the enclosure. X-ray beams must be treated with the greatest respect. Safety interlocks can fail, and should never be relied on for protection.

The generation of X-rays

In the laboratory, X-rays are usually produced by focusing a beam of electrons on to a metal target in a vacuum. The electrons are released from a cathode whose potential is tens of kilovolts below that of the metal anode, so that each electron possesses enough energy to produce an X-ray photon. The electrons cause transitions in the metal atoms of the anode, which release photons of particular energies (known as 'characteristic' radiation) as well as a general spectrum (known as 'white' radiation) (Fig. 1.14).

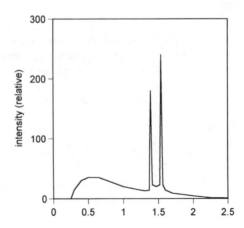

Fig. 1.14 X-ray spectrum generated by a copper anode bombarded with 50 keV electrons (schematic).

The X-rays used for X-ray crystallography usually have wavelengths between 1.6 Å and 0.5 Å. The wavelength is just shorter than the 'soft' X-ray region, and they penetrate biological materials adequately to be scattered from the whole volume of a crystal. In practice, laboratory X-ray generators usually produce characteristic radiation from a copper anode (CuKα, 1.54 Å wavelength). If shorter wavelengths are required, molybdenum may be used as an anode (MoKα, 0.71 Å).

In order to create an intense X-ray source, an electron beam from a heated wire at a potential of −30 kV or so is focused on a very small area of metal, and a large amount of heat is generated in a small volume. The rate at which this heat can be conducted away limits the intensity of the X-ray beam. Laboratory sources for the study of macromolecules must be as intense as possible, and the anode is usually a rotating copper cylinder, bombarded at a point on the outside by an electron beam, and cooled on the inside by jets of water. Such a source is a rotating-anode X-ray generator (Fig. 1.15).

Miniature X-ray sources have recently been designed which generate small but intense X-ray beams from a fixed anode, which may be focused so effectively by a shaped X-ray mirror that they compete with rotating-anode generators when small crystals are used. These may become the chosen laboratory sources for study of very small crystals (Fig. 1.16).

Since the 1960s, electron synchrotrons have become available as much more intense X-ray sources. These are large devices in which electrons travel on a circular track in a vacuum, influenced by electric fields which maintain the electrons at a velocity more than 99% of the speed of light (Fig. 1.17). As the electrons circle around the synchrotron, they

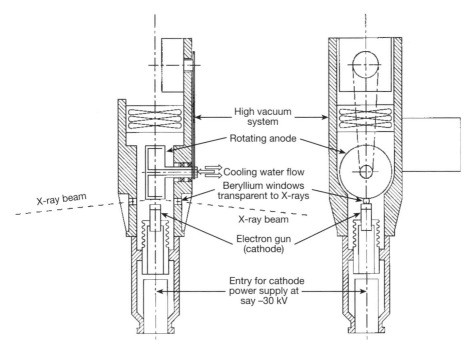

High vacuum system

Rotating anode

Cooling water flow

Beryllium windows transparent to X-rays

X-ray beam

X-ray beam

Electron gun (cathode)

Entry for cathode power supply at say −30 kV

Fig. 1.15 Schematic drawing of a rotating-anode X-ray generator (courtesy of Bruker Nonius B.V., Delft, The Netherlands).

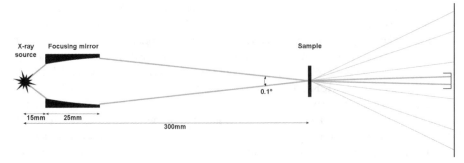

Fig. 1.16 A curved cylindrical mirror focuses X-rays from a microsource, giving high intensity in a very small area (courtesy of Bede Scientific Instruments, Bowburn, UK).

Fig. 1.17 The European Synchrotron Research Facility in Grenoble, France. Behind the circular beam hall, the high flux neutron beam reactor of the Institut Laue-Langevin may be seen. Its beam hall extends from the reactor dome towards the synchrotron. (Provided by Jean-René Regnard and Marie-Claude Moissenet.)

emit an intense X-ray beam in the tangential direction. Various electrical disturbances can be put in their path which modify the radiation that they produce. Synchrotron sources produce a continuous X-ray spectrum, making the whole useful range from 0.5 Å to 1.6 Å available for experiments.

How X-rays interact with matter

If an X-ray photon encounters an electron, it may be absorbed and sets the electron into vibration at the X-ray frequency. This vibrating electron then emits an X-ray photon in a random direction, of the original energy and wavelength. This is called 'coherent scattering'. More often, when the electron is in an atomic orbital, the interaction may cause transitions in the atom, resulting in emission of one or more photons of lower energy ('incoherent scattering'). This releases energy in the atom, and may leave it in an excited state. Radiation damage is a consequence of incoherent scattering, while crystallographic structure determination depends upon coherent scattering.

A photon can also interact with an atomic nucleus, but the nucleus is so much more massive that it vibrates much more weakly and its scattering of photons is negligible. Therefore X-ray crystallography and X-ray scattering experiments 'see' electrons, and the structural results represent distributions of electron density.

The properties of X-rays have an important effect on the way X-ray diffraction measurements are made

The X-rays used in crystallography are not very penetrating. The incident beam is almost all absorbed in a few millimetres of protein or water. It is no use making specimens as thick as this, since the beam is absorbed, and few X-rays will reach the detector. Depending on the wavelength used, the ideal size for a protein crystal is from 200 to 1000 μm (0.2–1 mm), giving a specimen which scatters a great deal of the incident beam, but does not absorb so much as to lose the signal. If the X-ray source is sufficiently intense, smaller crystals down to a few microns in size may be used. Since shorter wavelengths became available at synchrotrons, the advantages of moving from 1.5 Å X-rays to 0.9 or 0.6 Å have shown themselves. One important improvement is that corrections made for the absorption of X-rays in the specimen are much less important.

Crystals present molecular matter in a highly ordered form, in which (ideally) every molecule is precisely positioned in relation to its neighbours. But every time an X-ray photon is scattered incoherently, serious damage is done to the crystal. A simple approximation is that the molecule which scattered it is totally disrupted. Fortunately there are many molecules in a crystal (usually 10^{15} or more), and many photons can be scattered before the overall structure is destroyed.

If we tried to image one molecule with X-rays, it would be destroyed as soon as it scattered one or two photons. Even if we could build an X-ray microscope of adequate resolving power, an image could not be generated. This is one reason why crystals can be used for structural study, while attempts to image a single molecule would be hopeless. Since a crystal contains so many identical molecules, the radiation damage problem is not so severe. The crystal is not too badly damaged if a small fraction of its molecules are destroyed by X-rays.

Exactly the same problem arises in electron microscopy. It is possible to focus electrons, using an electric field to act as a lens. In fact, an electron causes less radiation damage than an X-ray of corresponding wavelength. But in making an electron microscope image of a single object, so much damage is done that fine detail of its structure is inevitably lost. Electron micrographs of biological materials rarely show details smaller than about 20 Å. To interpret finer details the images of thousands of molecules have to be put together.

The nature of crystallographic images

The fact that crystallographic images cannot be made by lenses or mirrors is a nuisance to the experimenter, but brings important advantages in the end.

In crystallographic studies, scattering by the crystal is observed in all possible directions. The X-ray beam is incident on the crystal in a wide range of directions, and as much as possible of the scattered radiation is recorded. This makes the resulting image truly three-dimensional, with no special difference in representation or accuracy for different directions. One atom or molecule cannot hide behind another.

Our eyes and brains are not like this, because our retinas are two-dimensional and see a scene from a particular point of view. This means we have to show each other two-dimensional images of the crystallographic results. Although representations may include a stereoscopic effect to help convey a third dimension, perception of it is limited in just the same way as our natural vision is limited (Fig. 1.18). But the results that these pictures try to portray are truly three-dimensional. These representations are considered in more detail in Chapter 11.

To understand crystallography and to appreciate its results, it is valuable to develop an ability to remember and imagine the details of three-dimensional scenes. Some effort is required to appreciate the underlying three-dimensional nature of an L-amino acid, a right-handed helix, or a Rossmann fold.

Fig. 1.18 Three-dimensional electron-density distribution, represented at a single level. The density represents a Trp-Glu-Ala-Pro sequence. This small fragment of density has been cut out of an electron-density map for a protein of over 500 residues.
(Reproduced from Britton *et al.* (2000) with permission from Elsevier Science.)

X-ray detectors

X-ray detectors respond to the energy of the diffracted beam, also known as its intensity, which (at a given wavelength) is proportional to the number of photons it delivers. The accurate measurement of diffracted X-ray intensities is the subject of Chapter 6, but the techniques are outlined below.

The classical X-ray detector, which led to the identification of X-rays in 1895 by Röntgen, is photographic film. In the photographic emulsion, a tiny crystal of silver iodide can absorb an X-ray photon which releases enough energy to create an imperfection in the crystal lattice. (With visible light, several photons are needed to create the same effect.) The process of development allows the silver iodide to be reduced to silver, blackening the emulsion, only in crystals which have an imperfection. Photographic film was generally used for protein crystallography up to the late 1980s, and is still often used for preliminary screening observations.

Other types of detector use the ionizations produced as X-rays pass through a gas. In the proportional counter, the ionizations allow a current to pass, proportional to the rate of production of ions. These counters give an accurate measure of X-ray intensity, but have no spatial resolution. If an absorber with a small hole is placed in front of it, the counter can detect X-rays scattered from a crystal into this hole. It can therefore only explore the diffraction one point at a time.

Diffractometers are devices which allow the crystal orientation and the position of the counter to be controlled accurately while the crystal is kept in a finely collimated X-ray beam. They are widely used for non-biological crystallography, where the diffraction pattern is much less complex but, because each individual measurement has to be made in sequence, they are laborious and slow for crystals of large molecules.

Several types of 'position-sensitive' detector have been devised which retain the digital accuracy of a counter, but allow a large area to be observed. The most successful in the 1990s was the *image plate* (Fig. 1.19), which stores the diffraction image in a temporary form in which atoms are promoted to an excited state when they absorb an X-ray photon. The image plate can then be 'read' by a fluorescence technique. The intensity of fluorescence is monitored as a light beam scans the plate, giving a measure which is accurately proportional to the total X-ray energy which has arrived at each point. The plate is then 'erased' by flooding it with light, which releases all residual fluorescent centres. The image plate behaves like a more accurate, reusable form of photographic film. Like

Fig. 1.19 X-ray diffraction recording set-up using an image plate (courtesy of Mar X ray Research, Evanston, Illinois, USA).

film, it preserves no record of the time of arrival of the photons, and a separate scanning operation is needed to record the results. For complex diffraction measurements from macromolecular crystals, this means that the exposure has to be suspended at frequent intervals, after a small part of the total diffraction data has been recorded.

Other detectors have been devised to obtain the digital accuracy of a counter, recording the arrival of each photon in real time, while exploring a large area of the diffraction pattern. These include position-sensitive *multi-wire detectors,* and solid-state devices such as the *charge-coupled device* (CCD) (Fig. 1.20) and the *silicon solid-state detector.* The CCD

Fig. 1.20 A CCD detector in a transparent case. X-ray photons produce scintillations on a fluorescent screen at the left. The light is conducted by fibre optics on to a charge-coupled device at the centre of the unit. (Courtesy of Mar Xray Research, Evanston, Illinois, USA.)

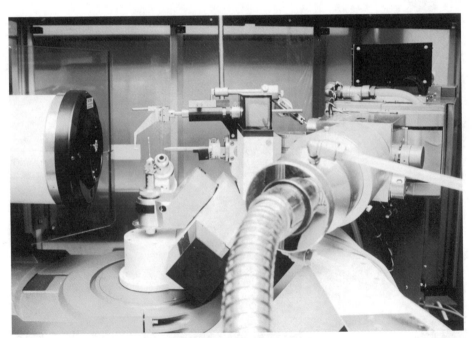

Fig. 1.21 A laboratory set-up using a Nonius rotating anode generator, crystal monochromator, and FAST detector. In the background part of the radiation enclosure can be seen.

is now the preferred detector for rapid observation, but is still limited in size. Another type of detector uses image-intensifying technology to produce an image which can be scanned by television-type methods to obtain a digital record of scintillations caused by individual photons. The FAST detector (Fig. 1.21) is an example of this type of device. These solid-state and image-intensifying devices provide more efficient detectors, though more limited in positional accuracy, than film or image plate. There are conflicting requirements for sensitivity, positional accuracy, and speed in all these devices.

Figure 1.21 shows a laboratory set-up for measuring scattered intensities using a FAST position-sensitive detector. At the right is a Nonius rotating-anode X-ray generator, with the electron gun nearest to the camera. This focuses a beam of high energy electrons onto the rotating anode in the box at the right. To its left is a cylindrical holder for an X-ray monochromator which allows only one wavelength to pass into the collimator tube. This tube, pointing towards the crystal at the centre, is accurately aligned with small pin-holes at each end. These pinholes ensure that the only X-rays reaching the crystal are almost parallel, coming from the small area of the generator's anode where the intensity is highest, as a fine beam only just larger than the crystal. The crystal is on a three-circle mounting which allows it to be rotated into any desired angular orientation. X-rays that pass through the crystal without being scattered are intercepted by the beam stop to its left. The only X-rays that can reach the circular detector at the extreme left are those which have been scattered by the crystal.

In the set-up shown, the X-ray beam reaching the detector has a significant distance to travel from the crystal. Especially at longer wavelengths (say 1.2 Å or more) the scattering of X-rays by air between crystal and detector can significantly degrade the observations. A shaped plastic balloon filled with helium, which scatters X-rays much less strongly, may be placed between the crystal and the detector.

Other types of wave

Diffraction experiments are possible using other types of incident 'rays'. The most important applications to biological materials use electrons and neutrons, and some electron diffraction techniques were mentioned above. A large neutron beam generator is shown in Fig. 1.17.

These applications exploit the quantum theory's demonstration that matter, like light, has two aspects to its existence—a wave-like quality as well as a particulate quality. Specimens scatter them as they scatter other types of wave. Some details are given in Box 1.2.

The use of electrons in an electron microscope has been mentioned above. Electron diffraction is particularly well suited to studying biological membranes. An appropriate electron beam is much less penetrating, and is strongly scattered even by a monomolecular layer. By making a thin crystalline sample, or an ordered monomolecular layer such as a two-dimensional crystal, images can be derived from electron scattering by techniques similar to those used in X-ray diffraction.

The properties of neutron scattering make it useful for special studies, particularly where hydrogen positions need to be observed. Neutrons cause almost no damage to the diffracting specimen. Neutrons are, however, comparatively expensive to produce, and larger crystal specimens are needed because neutrons interact more weakly with matter.

Box 1.2 Matter waves

Other sorts of waves exist, which can have appropriate wavelength. The two most important for structure determination are the waves associated with electrons and with neutrons.

Electrons have the advantage of being easily focused, but have little penetrating power and must be used on very thin specimens in a vacuum. The focusing properties are exploited in the electron microscope, which can produce direct images of metal lattices at atomic resolution. For biological specimens, due to the damage caused by the electron beam, a single particle cannot be imaged at much better than 25 Å resolution.

Modern electron diffraction techniques use very small crystals, distributing the damage over all the particles in the array, and using methods very like those of crystallography to overcome the problems of radiation damage. The weak penetration is exploited conveniently in studying membranes, where an image may be generated from a two-dimensional crystalline layer of a membrane protein.

Neutrons are very penetrating, and require large crystals for convenient observation. They cause negligible radiation damage and all neutron scattering is coherent scattering. The neutron scattering power depends on the properties of the nucleus and is not a simple function of the atomic weight. Hydrogen scatters neutrons strongly, but very differently from deuterium. Crystal neutron diffraction is especially useful to determine hydrogen positions, and to investigate hydrogen exchange with the solvent.

The wavelength associated with a massive particle is given by the de Broglie relation:

$$\lambda = h/p,$$

where p is the momentum of the particle, and h is Planck's constant.

An electron with an energy of about 150 eV has a wavelength of 1 Å. More energetic electrons (using larger voltages to produce shorter wavelengths) are used in practice, to increase their penetrating power and for convenient microscope design.

A neutron, being much heavier, requires much lower energy for a wavelength of 1 Å. The mean thermal energy of any particle at temperature T is $(3/2)kT$ ($(1/2)kT$ in any particular direction). A neutron travelling at about 5 km/s has a wavelength of about 1 Å. This velocity is the mean velocity of neutrons at 635 K.

In practice, less energetic neutrons are often used, with somewhat longer wavelength. Neutrons equilibrated to room temperature before emission from a nuclear reactor are often used, with a wavelength of about 1.5 Å, travelling at about 2.7 km/s.

Further reading

A few textbooks are aimed specifically at protein crystallography:
Blundell, T.L. and Johnson, L.N. (1976). *Protein crystallography*. Academic Press, New York.
Drenth, J. (1994). *Principles of protein X-ray crystallography*. Springer, New York.

Rhodes, G. (2000). *Crystallography made crystal clear* (2nd edn). Academic Press, San Diego.

Many textbooks of crystallography present crystallographic methods and theory in detail. The following may be more appropriate to readers from a biochemical background:
Glusker, J.P. (1994). *Crystal structure analysis for chemists and biologists*. VCH, New York.

Ladd, M.F.C. and Palmer, R.A. (1993). *Structure determination by X-ray crystallography* (3rd edn). Plenum Press, New York.

Stout, G.H. and Jensen, L.H. (1989). *X-ray structure determination: a practical guide* (2nd edn). Wiley, New York.

A compact and readable general text on crystallography:
Clegg, W. (1997). *Crystal structure determination*. Oxford University Press, Oxford.

A more complete survey of crystallography, including macromolecular aspects:
Giacovazzo, C. *et al.* (1992). *Fundamentals of crystallography*. International Union of Crystallography/Oxford University Press, Oxford.

A historic book with special relationship to protein crystallography was written by Sir Lawrence Bragg, but required final revision after his death in 1971. Bragg was a wonderful teacher and an influential supporter of the early development of protein crystallography.
Bragg, W.L. (1975). *The development of X-ray analysis* (ed. D.C. Phillips and H. Lipson). Bell, London.

The early history of stereochemistry:
Traynor, J.G. (ed.) (1987). *Essays on the history of organic chemistry*. Louisiana State University Press, Baton Rouge.

Wotiz, J.H. (ed.) (1993). *The Kekulé riddle*. Cache River Press, Vienna, Illinois.

The early history of X-ray crystallography:
Ewald, P.P. (ed.) (1963). *Fifty years of X-ray diffraction*. International Union of Crystallography/Oosthoek's Uitgeversmaatschappij, Utrecht.

History of macromolecular crystallography:
Rossmann, M.G. (2001). Historical background. In *International tables for crystallography*. Vol F. (M.G. Rossmann and E. Arnold, ed.) pp. 4–9. International Union of Crystallography/Kluwer, Dordrecht.

The physical properties of X-rays:
Als-Nielsen, J. and McMorrow, D. (2001). *Elements of modern X-ray physics*, Chapters 1 and 2. Wiley, New York.

X-ray optics and detectors (a brief survey with references):
Drenth, J. (1994). *Principles of protein X-ray crystallography*, Chapter 2. Springer, New York.

Leslie, A.G.W. (2000). In *Structure and dynamics of biomolecules* (ed. E. Fanchon *et al.*), pp. 14–20. Oxford University Press, Oxford.

2

Crystals and symmetry

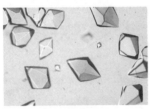

Fig. 2.1 Crystals of xylose isomerase from *Arthrobacter*.

One of the fascinations of crystallography is the beautiful external appearance of crystals (Fig. 2.1), and the corresponding beauty of the atomic arrangements within them. At an early stage of studying a crystal, a crystallographer needs to analyse its underlying symmetry.

This must be done because the crystallographic results must satisfy this symmetry and are constrained by it. It is needed to decide on the appropriate strategy for observation of X-ray scattering by the crystal. It is also essential to know the precise symmetry when interpreting the scattering data to obtain the crystal structure. The analysis of many crystal structures has been delayed because of mistakes in the symmetry assignment, and some have been incorrectly analysed.

This chapter sets out to give an overview of crystal symmetry. It does not provide an exhaustive presentation. In particular, we shall concentrate on the symmetrical arrangements of chiral objects, in which mirror symmetry is forbidden. Much more detail about crystal symmetry can be found in many textbooks of crystallography, and a complete reference guide is provided in *International Tables for Crystallography, Volume A* (see Further Reading at the end of this chapter).

What is symmetry?

An object is symmetrical if, after some operation has been carried out, the result is indistinguishable from the original object. Consider, for example, an equilateral triangle. If the triangle is rotated 120° about its centre, the resulting triangle is, in all respects, identical to the original triangle (Fig. 2.2). This means that a second 120° rotation, producing a total rotation of 240°, also makes no change to the object. A third 120° rotation makes, of course, a 360° rotation, bringing the triangle back to its original orientation. So these rotations are called *3-fold rotation operations*, and the equilateral triangle is said to possess *3-fold symmetry*.

The meaning of 3-fold symmetry is that the object may be rotated repeatedly by $360°/3 = 120°$ about its symmetry axis without changing it. Similarly 2-fold symmetry refers to rotation by $360°/2 = 180°$ and 6-fold symmetry refers to rotation by 60°.

A left hand has no symmetry. Reflect it in a mirror and what you see looks like a right hand. But many everyday implements possess mirror symmetry: consider, for example, a fork or a cup (Figs 2.3, 2.4). If we had only one half of the fork, and looked at it with the mirror in the position shown, the reflection precisely generates the missing half. This means, incidentally, that it is equally useful to left- and right-handed people. The *mirror plane* of symmetry for these objects is a plane passing through its centre. The symmetry operation is called *reflection*. Unless there are special decorations on it, the whole fork or cup looks identical when viewed in a mirror.

A pair of scissors has no mirror symmetry. Scissors which can be used effectively in the left hand are different from ordinary right-handed scissors. Some types of scissors do possess 2-fold symmetry (Fig. 2.5). They are made of two identical blades, and there is a direction about which they can be rotated 180°, interchanging one blade with the other,

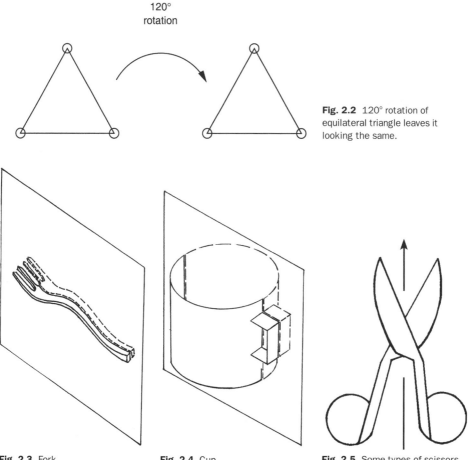

120°
rotation

Fig. 2.2 120° rotation of equilateral triangle leaves it looking the same.

Fig. 2.3 Fork.

Fig. 2.4 Cup.

Fig. 2.5 Some types of scissors do possess 2-fold symmetry.

leaving the pair of scissors apparently unchanged. This direction is called the *2-fold symmetry axis* of the object.

In these last examples, the symmetry is obviously imperfect. There is bound to be a scratch or a flaw which distinguishes one aspect of the object from the other 'almost-symmetrical' aspect. The scissors will have a screw or rivet holding the two blades together, which is probably different at its two ends, contradicting the symmetry. True symmetry applies only to ideal objects like an equilateral triangle.

Another kind of symmetry is called *centrosymmetry*. The symmetry operation here is to move every point in the object along the line joining it to a chosen centre point, and to continue along this line until it is equally far from the centre on the other side. This operation is called *inversion*. If such an operation is carried out on a rectangular block, the original shape may be reproduced (Fig. 2.6). An (ideal) rectangular block is said to be *centrosymmetric*. It may seem surprising at first, but a box with sides that are not rect-angles (a parallelepiped) is also centrosymmetric (Fig. 2.7). An object such as a hand or a foot, a glove or a shoe has no centre of inversion, but it is possible to arrange a pair of them (right and left) in a centrosymmetric fashion (Fig. 2.8).

We can be sure that biological macromolecules are not centrosymmetric. Natural proteins are made of L-amino acids, which the inversion operation would convert into D-amino acids (Fig. 2.9). There are similar asymmetric centres in sugars which ensure that crystals of sugars, polysaccharides or nucleic acids are not centrosymmetric.

It follows that no biological macromolecule has mirror symmetry. If you look at yourself in a mirror, the hair of the person you see has hair parted on the other side of the head. If you are right-handed, the person you see will be using the left hand. The image is an *inverted* version of yourself (using inversion in the sense defined above, as a sym-metry operation).

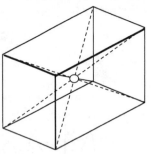

The types of symmetry operation for finite three-dimensional bodies are:

- rotation
- reflection
- inversion

Fig. 2.6 A rectangular box is centrosymmetric about its centre.

and only the first can exist in biological macromolecules, which lack a centre of symmetry and are called *chiral*.

Rotational symmetry operations must always be through an angle which is an integral fraction of 360°. They must be *n-fold symmetry operations*, where *n* is an integer.

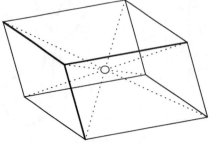

Fig. 2.7 A parallelepiped is also centrosymmetric about its centre.

Fig. 2.8 Inversion symmetry of a pair of hands.

Otherwise, repeated operation of the symmetry never gets back to the original object, so it would generate an infinite number of units.

Symmetry of molecular oligomers

Many protein molecules are composed of several identical peptide chains in a symmetrical arrangement. Protein dimers usually have 2-fold rotational symmetry (about an axis called the dimer axis), and trimers usually have 3-fold rotational symmetry about a trimer axis (Figs 2.10 and 2.11).

Many proteins exist as symmetrical tetramers. Although it is possible for them to have 4-fold rotational symmetry (Fig. 2.12), this is relatively unusual. More often, the tetramer has 2-fold symmetry about each of three perpendicular directions. To see how this is built up, take the dimer with 2-fold symmetry shown in Fig. 2.10 and set up another perpendicular 2-fold axis which intersects the first. This will generate two more copies of the chiral unit (Fig. 2.13). The tetramer created in this way has a third 2-fold axis which has been generated spontaneously. This third 2-fold axis is perpendicular to each of the others, and all three axes meet at a single point. One monomer is related to each of the other three monomers by a 2-fold rotation about one of these three axes. This type of symmetry is called 222 symmetry because it has three 2-fold axes. Figures 2.14 and 2.15 view this object down 2-fold axes. Notice that the three 2-fold axes are quite different. One passes near the wrists, one near the thumbs, one near the little fingers. This arrangement is found, for example, in the symmetrical tetramer of glyceraldehyde 3-phosphate dehydrogenase (Fig. 2.16).

The kinds of symmetry that can be possessed by a local assembly of objects are called the *point groups*. The point group named 222 identifies three types of symmetry operations which relate the subunits of glyceraldehyde 3-phosphate dehydrogenase. A 2-fold rotation about the x-axis, a 2-fold rotation about the y-axis or a 2-fold rotation about the z-axis are all symmetry operations of this arrangement.

By creating a 2-fold symmetry axis perpendicular to any n-fold symmetry axis, a second kind of 2-fold axis is always generated. If a 2-fold axis is introduced close to the thumbs in these diagrams, a second is generated close to the little fingers. For every point group with n-fold symmetry, another exists with $n22$ symmetry. Figure 2.17 shows the 7-fold symmetry of a chaperonin.

We shall see later that crystals can only accommodate certain kinds of symmetry, because of the constraints of the crystal lattice. The only rotational symmetries which a crystal can have are 2-, 3-, 4-, and 6-fold symmetry. There are thus a limited number of so-called *crystallographic point groups*. There are 11 crystallographic point groups for chiral units, as shown in Fig. 2.18.

The last two of these are a bit different from the others, because they have the 3-fold symmetry that you will see if you view a cube from one corner. These *cubic* point groups

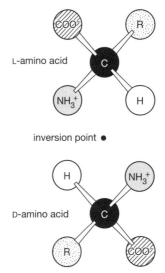

Fig. 2.9 An L-amino acid inverted through the central point makes a D-amino acid.

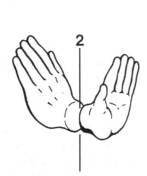

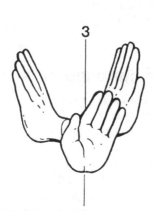

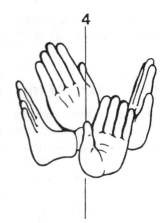

Fig. 2.10 Symmetrical dimer of a chiral object. The 2-fold rotational symmetry axis is indicated.

Fig. 2.11 A symmetrical trimer, showing its 3-fold axis.

Fig. 2.12 A 4-fold symmetrical tetramer.

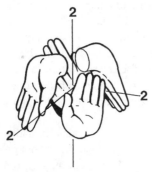

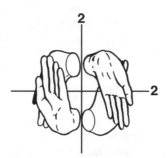

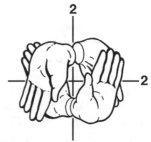

Fig. 2.13 A symmetrical tetramer with 222 symmetry.

Fig. 2.14 The 222 tetramer viewed along one of its 2-fold axes.

Fig. 2.15 The 222 tetramer viewed along a different 2-fold axis.

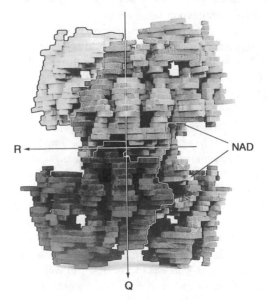

Fig. 2.16 A low-resolution model of glyceraldehyde 3-phosphate dehydrogenase. The four protein monomers are identical in shape, but have been given different shading. (Courtesy of Alan Wonacott.)

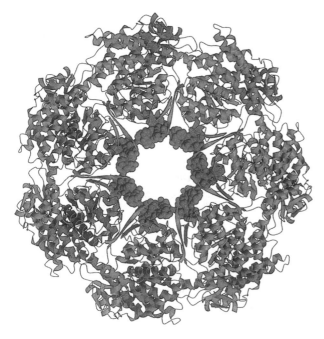

Fig. 2.17 In its unliganded form, GroEL has 7-fold symmetry. (Reprinted from Chen and Sigler (1999) with permission from Elsevier Science.)

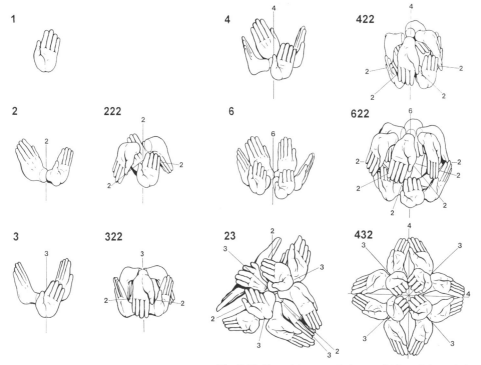

Fig. 2.18 The point groups that can exist in protein crystals.

have 3-fold axes in four different directions. The simpler one, called 23, also has 2-fold axes in three perpendicular directions. The other, 432, has 4-fold axes in these directions and 2-fold axes at 45° to them.

This would have been familiar to Plato, who identified the regular three-dimensional solids—the tetrahedron, the cube, the octahedron, the dodecahedron, and the icosahedron. All of these have four 3-fold axes parallel to the body diagonals of a cube.

There is one more cubic point group, which is a non-crystallographic point group because it includes 5-fold axes. Highly symmetrical protein oligomers are found in the capsid of many spherical viruses. This packing provides an effective way for the capsid to enclose a large amount of nucleic acid, using a relatively small capsid protein molecule. These capsids have the symmetry associated with the most complex of the Platonic solids, the 20-sided icosahedron. This symmetry has become familiar because footballs are made and decorated in a way that exploits it (Fig. 2.19). There are 5-fold symmetry axes at the centre of each pentagon, 3-fold axes relating three adjacent pentagons, and 2-fold axes at the corners of each pentagon. The symmetry of these viral capsids is called 532. The 5-fold symmetry applies to the individual virion. As explained previously, this symmetry cannot be expressed in a crystal of icosahedral virus particles.

Many symmetrical viral capsids are formed of 60 similar units. The operation of icosahedral or 532 symmetry on the monomeric coat protein generates 60 copies of it (Fig. 2.20).

It is also possible for a larger aggregate, such as a trimer, to be arranged so that the aggregates are arranged with 532 symmetry. Although the 532 symmetry can be exact, any higher symmetry is only approximate. This is an example of *pseudo-symmetry*, which is a common occurrence in biological molecular structures. Capsids of spherical viruses composed of 180 monomer units are common (Fig. 2.21).

Crystals

Crystals present matter in its most highly ordered form. This high degree of order provides the special possibility of imaging the molecules, since (ideally) all the units that build up the crystal are completely identical, and have an identical environment. This opportunity of imaging crystals is the central concern of this book.

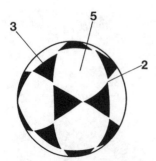

Fig. 2.19 Footballs are often decorated in a way that shows 532 symmetry.

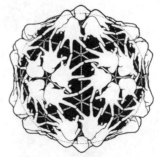

Fig. 2.20 Fanciful drawing of left hands arranged in 532 symmetry by Don Caspar (reproduced from Caspar (1980) by permission of the Biophysical Society).

Fig. 2.21 Pseudo-symmetrical arrangement of 180 units (reproduced from Harrison (1980) by permission of the Biophysical Society).

Crystals may be composed of atoms, ions, or molecules. For our purposes, the primary interest is in molecular crystals. They are built up in a regular way so that equivalent molecules each interact identically with their neighbours (Fig. 2.22). Ignoring the limitation imposed by the finite size of the crystal, each repeating unit within the crystal has an identical environment. The way the molecules stack together allows a crystal to have flat faces with sharp edges (Fig. 2.23), although crystals do not always grow like that. When molecules are big enough to be seen in the electron microscope, this feature of a crystal can be visualized directly (Fig. 2.24).

Fig. 2.22 In an ideal crystal, the contents of every unit cell are identical. Each unit interacts identically with its neighbours.

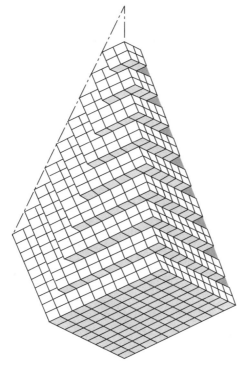

Fig. 2.23 The stacking of unit cells allows a crystal to have flat faces with sharp edges (after Haüy 1801).

Crystals grow if the solution that surrounds them is more than saturated. This idea is familiar enough with salt solutions, but applies equally to protein solutions. Proteins being polyions, their solubility is affected by the concentration of other charged ions, including other polyelectrolytes. The presence of ammonium sulphate, of other ionic compounds, or of polymers which sequester water, such as polyethylene glycol (PEG), reduces protein solubility and favours crystal growth.

When a protein crystal grows from an oversaturated solution, its molecules are deposited on the surface of the growing crystal, and come into exactly the correct orientation to continue the pattern of crystal growth. Ideally, if a molecule is wrongly oriented, or if it is the wrong kind of molecule, it will dissociate from the crystal surface before it becomes locked into place by adjacent molecules settling on the crystal surface.

In macromolecular crystals, since the irregularly shaped molecules cannot pack so as to fill all space, there will be voids between the molecules, and these are filled by the aqueous growth medium (Fig. 2.22). Interaction between the molecules is over a small area, making the crystals soft and fragile. The aqueous inclusions are usually in a liquid-like state. In practice, the starting material will include impurities, some very similar to the molecule which is being crystallized, but slightly damaged, and other much smaller molecules. Some of these impurities will get trapped in the crystal. These factors make macromolecular crystals generally less perfect than ionic crystals.

The crystal lattice

Crystals form an extended three-dimensional array of molecules which build up a new type of symmetry. In a previous section, a distinction was drawn between the approximate symmetry of real objects, and the perfect symmetry of ideal objects. Macromolecular crystals are usually far from perfect, with consequences for all aspects of their properties. Remembering that our real crystals have many imperfections, it is necessary to imagine an ideal crystal, in order to study its symmetry. This ideal crystal will also be assumed to be infinite in size!

An ideal crystal has a type of symmetry known as *lattice symmetry*. A two-dimensional lattice is shown in Fig. 2.25. If it is moved so that one of the repeating units of its structure takes exactly the position formerly occupied by a neighbouring one, the resulting object is indistinguishable from the original crystal (Fig. 2.26). (If the crystal were finite, its boundaries would move, but the boundaries of an infinite crystal are inaccessible!)

Fig. 2.24 Electron micrograph of poliovirus.

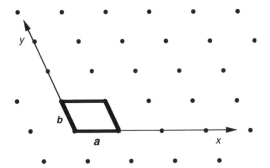

Fig. 2.25 A two-dimensional lattice.

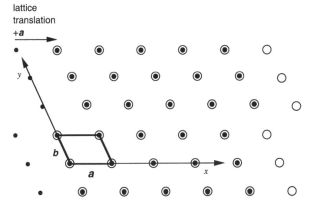

Fig. 2.26 The lattice of black dots has been displaced by $+a$ and shown as white circles. The lattice point positions are unchanged, but the boundary has moved. If the lattice were infinite, no change could be observed.

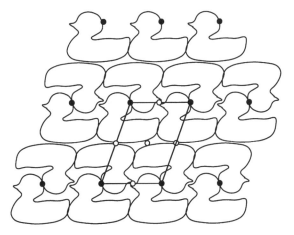

Fig. 2.27 If there is 2-fold symmetry about the lattice points of a two-dimensional lattice (black circles), the lattice translations generate new types of 2-fold symmetry (open circles), so that four different types of 2-fold axis exist.

This operation is called a *lattice translation*. Since a lattice translation is a symmetry operation, any combination of successive lattice translations is also a symmetry operation.

Staying for the moment in two dimensions, it may be observed that rotational symmetry may be added to the lattice symmetry. Figure 2.27 shows a two-dimensional lattice with 2-fold symmetry. One 2-fold axis is chosen as origin, but the lattice symmetry generates three different types of 2-fold symmetry axis, which have different environments. If there is 3-fold symmetry (Fig. 2.28) three different types of 3-fold symmetry axis exist.

The unit cell and space group

Three lattice translations can define a three-dimensional crystal *unit cell*, a solid block whose edges correspond to the three lattice translations (Fig. 2.29). The lattice translations can have different lengths and may be at arbitrary angles (but as we shall see, symmetry restricts the possible angles). Box 2.1 states conventions which apply in choosing these axes.

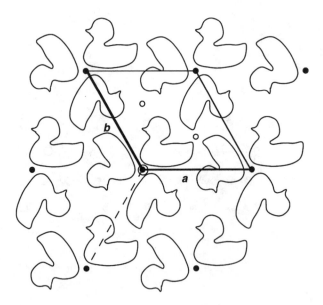

Fig. 2.28 If there is 3-fold symmetry, the lattice is generated by two lattice translations which make an angle of 120° and are of equal length. When objects are arranged with 3-fold symmetry about the lattice points, two other types of 3-fold symmetry axis are generated, indicated within the outlined cell.

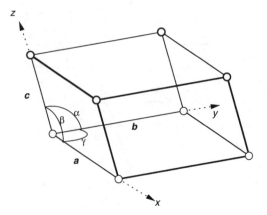

Fig. 2.29 The unit cell dimensions are specified by three lattice translations whose lengths are *a*, *b*, and *c*. The angles between them are α, β, and γ. When written in bold type *a*, *b*, *c* specify both the lengths and the directions of the unit cell edges. The coordinates *x*, *y*, and *z* are measured along *a*, *b*, and *c* respectively.

Positions in the unit cell are defined by distances x, y, and z representing fractional displacements along these three axial directions (Box 2.2). It is usual to define a unit cell which has the smallest possible volume. It is the smallest unit which can build up the whole crystal by lattice symmetry operations. A unit cell with the smallest possible volume is called a *primitive unit cell*, and its edges define *primitive lattice translations*. The entire lattice is composed of unit cells stacked in three dimensions, each with the same orientation (Figs 2.23, 2.30). In an ideal crystal, the contents of every unit cell are identical, as in the 2-dimensional example of Fig. 2.22.

Lattices can have the same types of symmetry that other objects can have, but rotational symmetry imposes restrictions on the shape of the unit cell. Since the stack of unit cells has to fill space and fit the rotation, the lattice translations must fulfil the symmetry. If there is 2-fold symmetry, there must be two lattice translations which are perpendicular to the symmetry axis (Fig. 2.31). Three-fold (Fig. 2.32), 4-fold, and 6-fold symmetry make similar requirements: one lattice translation must be parallel to the rotation axis, and perpendicular to two other lattice translations. The requirements are summarized in Table 2.1, and Box 2.1 gives more detail.

A less obvious fact, already alluded to, is that the only rotational symmetries a lattice can have are 2-fold, 3-fold, 4-fold, and 6-fold. It is not possible to fill a plane with regular pentagons, and no lattice with 5-fold symmetry can exist. The eighteenth-century mathematician Leonhardt Euler showed this to be also true for all rotational symmetries 7-fold and

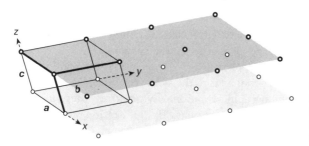

Fig. 2.30 A small part of a three-dimensional lattice. Two layers of the **a**–**b** plane are shown, and lattice points in the upper layer are enhanced.

Table 2.1 The crystal systems

Crystal system	Minimum symmetry requirement	Conventional choice of axes	Constraints on interaxial angles and axial lengths
Triclinic	None	No constraints	None
Monoclinic	One 2-fold axis	**b** parallel to 2-fold **a** and **c** perpendicular to 2-fold	α and $\gamma = 90°$
Orthorhombic	Three perpendicular 2-fold axes	**a**, **b**, and **c** parallel to 2-fold axes	α, β, and γ all 90°
Trigonal*	One 3-fold axis	**c** parallel to 3-fold **a** and **b** perpendicular to 3-fold	$\beta = 120°$, α and $\gamma = 90°$ **a** and **b** equal length
Tetragonal	One 4-fold aixs	**c** parallel to 4-fold **a** and **b** perpendicular to 4-fold	α, β, and γ all 90° **a** and **b** equal length
Hexagonal	One 6-fold axis	**c** parallel to 6-fold **a** and **b** perpendicular to 6-fold	$\beta = 120°$, α and $\gamma = 90°$ **a** and **b** equal length
Cubic	Four 3-fold axes	**a**, **b**, and **c** related by 3-fold axis	α, β, and γ all 90° **a**, **b** and **c** equal length

*A special case of the trigonal system, known as the rhombohedral system, is not discussed here.

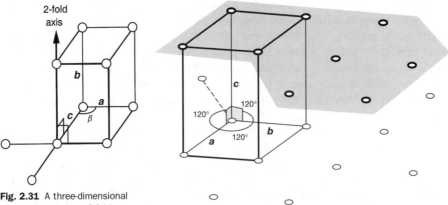

Fig. 2.31 A three-dimensional lattice can only have 2-fold symmetry if two of the lattice translations are perpendicular to the 2-fold axis.

Fig. 2.32 Two layers of a three-dimensional lattice with 3-fold symmetry, showing the **c** translation perpendicular to **a** and **b**. Symmetry makes a third lattice translation (dashed) equivalent to **a** and **b**.

Box 2.1 Conventions about choosing crystal axes

In the case of 2-fold symmetry, the 2-fold axis must be perpendicular to the two other lattice translations (Fig. 2.31). This is necessary to allow the lattice translations to obey the 2-fold symmetry. These two lattice translations define a lattice plane. Thus, the 2-fold symmetry has to be perpendicular to a lattice plane.

You can easily see that the same applies to a 3-, 4-, or 6-fold axis, and that again there has to be a lattice plane perpendicular to the symmetry axis (Fig. 2.32).

Thus, for all possible crystallographic rotational symmetries, there is always a lattice plane perpendicular to the rotation axis. By convention, we choose the crystallographic axes so that they show off the lattice symmetry. The highest-order symmetry axis is always chosen as a lattice translation direction, and the relation of the other symmetry axes to this axis defines the choice of the other two axes, which will always be related to each other by operation of the highest-order rotation axis, whether it is 2-fold, 3-fold, 4-fold, or 6-fold.

These conventional choices of crystal axes classify the possible crystal symmetries into *crystal systems*, listed in Table 2.1. Lattice translations along the axial directions are named *a*, *b*, and *c*, and there are some conventions about the choice of names. Subject to these conventions, the axes are chosen to give a unit cell of the smallest possible volume.

The crystal unit cell (Fig. 2.29) represents a building block, such that the complete crystal structure can be built up by stacking these blocks. In addition to the lengths of edges *a*, *b*, and *c*, the angles α, β, γ between the edges must be defined. According to convention, the lattice translations are chosen so that α, β, and γ are obtuse angles, but near to 90° if possible. Table 2.1 gives the relations between the crystal axes and axial angles, which are required in different crystal systems.

The names of some of the crystal systems look weird, but they are logical. Thus the *triclinic* system requires three arbitrary angles α, β, γ to be defined, but the *monoclinic* system has only one arbitrary or inclined angle, since the other two must be 90°. The unit cell of the *orthorhombic* system is a rhomb shape which has all its edges at 90°, that is, orthogonal.

higher. These symmetries are therefore not possible components of lattice symmetry, although, as already shown, a single assembly of molecules can have these symmetries.

Occasionally it is not possible to choose a primitive cell with axes parallel to the conventional crystal axes (Fig. 2.34). In these cases, by convention a compound unit cell is chosen with a volume two or more times the volume of the primitive cell. These choices are described in Box 2.3. The symbol P for a primitive unit cell is replaced by another letter (C, F, I, or R). One such lattice symmetry (C2) will be presented in detail.

According to the symmetry of the crystal lattice, the conventional unit cell may often contain several equivalent units (imagine them as molecules, but they could be other assemblies), some of them rotated relative to others by symmetry operations. The whole symmetry of the crystalline lattice is defined by its *space group*. This is the complete group of crystal symmetry operations that generate the three-dimensional lattice and define its symmetry. It will include three lattice translations, and the other operations that relate different units within the cell.

For example, any unit cell with a 2-fold symmetry axis contains at least two equivalent units forming a symmetrical dimer (Fig. 2.35). The smallest unit of the crystal that can generate the complete crystal structure by means of its symmetry operations is called the *crystal asymmetric unit*. It is called the asymmetric unit because it needs no symmetry of its own. Figure 2.36 provides a two-dimensional example. In a crystal there can be many possible ways of choosing the asymmetric unit, but every one has the same volume.

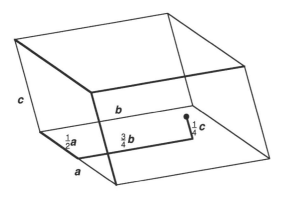

Fig. 2.33 The point $x = 1/2$, $y = 3/4$, $z = 1/4$, which may also be written (1/2, 3/4, 1/4).

Box 2.2 Positional co-ordinates

The three lattice translations defining a unit cell are called a, b, and c. When written in bold type, these quantities define a direction as well as a distance. The positive directions of the axes are chosen to make a right-handed set (such that a right thumb, forefinger and middle finger can be pointed along $+a$, $+b$, and $+c$, respectively). The three axes can never lie in the same plane.

Any point in the unit cell may be identified by a measurement along the three axial directions as x, y, and z. It is important to remember that x, y, and z measure distances as fractions of the unit cell dimensions. Thus $x = 1/2$ means a distance of half the unit cell along a, a distance which is $a/2$. Figure 2.33 shows the point $x = 1/2$, $y = 3/4$, $z = 1/4$ in an oblique lattice.

Fig. 2.34 A monoclinic lattice, with 2-fold symmetry axes parallel to **b**, face-centred on the rectangular face of the unit cell containing **a** and **b** (known as a C2 lattice). A primitive cell (dashed) has no rectangular faces. The 'conventional' face-centred cell has **a** and **c** axes perpendicular to **b**, but there are two lattice points per unit cell.

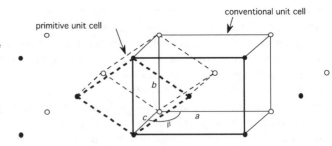

Fig. 2.35 A symmetrical dimer.

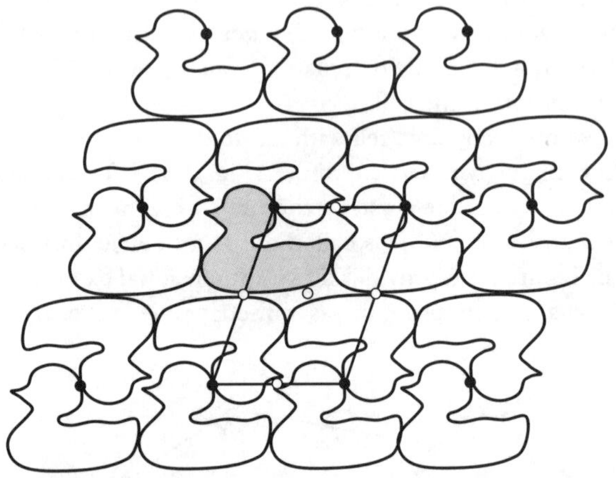

Fig. 2.36 The smallest unit of the structure that can generate the complete crystal structure by means of its symmetry operations is called the crystal asymmetric unit.

Box 2.3 Conventions about choosing unit cells

The convention is to choose unit cells which show off any rotational symmetry, and the origin is usually chosen so that symmetry axes pass through it. Subject to this convention, the unit cell is chosen to have the smallest possible volume.

In some cases, although the lattice symmetry enforces perpendicular lattice translations (Table 2.1), the choice of these as unit cell dimensions creates a unit cell larger than the minimum volume. Figure 2.34 gives an example. A *primitive unit cell* of minimum volume may be chosen, identifying a *primitive* (P) lattice, but the lattice translations of this cell are not perpendicular.

In these cases, perpendicular directions are chosen for the lattice translations, defining the *conventional unit cell*, which then contains more than one lattice point. With this choice the lattices may be found to be *face-centred* or *body-centred*; that is to say, lattice points are located at the corners of the cell (as usual), but also at a position in the centre of one face (C), at the centre of all the faces (F), or at the centre of the cell (I). A rhombohedral cell (R) is also possible in trigonal crystals.

Sodium chloride (Fig. 1.11) provides a familiar example of a lattice with all faces centred (F). If a cube-shaped unit cell is chosen with a sodium ion at each corner, there is an equivalent sodium at the centre of each face, (coordinates $x = 1/2$, $y = 1/2$, $z = 1/2$). A space group with one centred face, C2, is presented in detail in Box 2.6.

Positional coordinates x, y, and z are used to define the position of a feature in the unit cell with respect to the crystal axes (Fig. 2.33, Box 2.2).

The complete group of symmetry operations that apply in a particular crystal lattice is known as its *space group*. The concept of a space group can be made clearer by working through the examples in the following sections.

The simplest three-dimensional space group: space group P1

Space group P1 is the simplest space group, since it has no symmetry except the crystal lattice translations. In the symbol P1, the P indicates a primitive unit cell (rather than face-centred or body-centred), and the highest rotational symmetry is 1-fold (that is, only a rotation of 360°, which in practical terms means no rotational symmetry at all). The symmetry of this space group, therefore, is all in the lattice translations, which are labelled **a**, **b**, and **c**. These three translations form the edges of the unit cell.

Many different lattice translations might be chosen, relating different lattice points. Obviously they must not lie in the same plane, because the lattice must extend into a third dimension. Also, they are chosen to define a unit cell of the smallest possible volume. In this lowest-symmetry case, the unit cell contains only one asymmetric unit, and the lattice symmetry extends this volume to fill the whole space with similar units.

There are still many possible choices for the crystal axes, which may make any angles with each other. As already illustrated, the unit cell shape is a parallelepiped, shaped like a squashed brick (Fig. 2.29). There are three arbitrary angles specifying the unit cell, so these crystals are called *triclinic* (meaning three inclined directions). Since there is no symmetry within the unit cell, the whole cell forms the asymmetric unit. There are no particular 'landmarks' in the lattice, such as might be created by symmetry axes, and so the *origin* of the unit cell may be placed wherever is convenient. (The origin is the reference point for measurement of positions as x, y, z coordinates along the axes **a**, **b**, and **c**.) These details are summarized in Box 2.4.

Box 2.4 Space group P1

The unit cell is chosen by assigning three translation vectors, **a**, **b**, **c**, which are not in the same plane and define a minimum volume. The coordinate axes x, y, z are measured along the axes **a**, **b**, **c** as a fraction of the appropriate unit cell translation. Conventionally, the unit cell is chosen so that the axes make angles with one another which are as near to 90° as possible, subject to defining the minimum volume, and with the positive directions of **a**, **b**, and **c** making obtuse angles (greater than 90°) with each other.

In P1, any point in the unit cell has no equivalent point in the same cell. The choice of origin is an arbitrary one.

Space group P2

The next space group to consider is space group P2. In the symbol P2, the P indicates a primitive lattice, and the 2 indicates 2-fold symmetry. In space group P2, the lattice creates a system of parallel 2-fold symmetry axes, and the lattice translation parallel to the 2-fold axis is conventionally assigned as **b**. The molecular arrangement is illustrated in Fig. 2.37. The lattice directions **a** and **c** must be perpendicular to **b**. There is now only one arbitrary angle (β) to specify the unit cell shape, and the crystal lattice is called *monoclinic* (one inclined angle). This is shown in Fig. 2.38 and more details are given in Box 2.5.

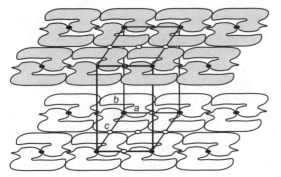

Fig. 2.37 Symmetry and equivalent positions in space group P2. A 2-fold axis along **b** creates two asymmetric units in the unit cell. Each unit has four 2-fold axes associated with it, at $x,z = (0,0)$ (black circles), and at $(0,1/2), (1/2,0), (1/2,1/2)$ (open circles).

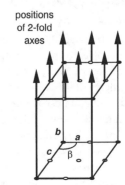

positions of 2-fold axes

Fig. 2.38 A unit cell of space group P2.

Box 2.5 Space group P2

One edge of the unit cell must be the lattice translation along the 2-fold axis direction, and conventionally this is denoted **b**. **a** and **c** are two lattice translations perpendicular to this which give a cell of minimum volume (Fig. 2.31). Conventionally the axes are chosen to make an angle near to 90° if possible, with an obtuse angle between the positive directions of **a** and **c**.

The origin of coordinates is chosen to lie on a 2-fold axis. This fixes a line along an axis where $x = 0$, $z = 0$, but the choice of origin along **b** is arbitrary. The 2-fold symmetry causes any feature at (x, y, z) to have an equivalent feature at $(-x, y, -z)$. Symmetry makes these two positions equivalent: they are known as *equivalent positions*. For each unit cell volume there are two equivalent asymmetric units.

The lattice symmetry creates many other equivalent positions including $(-x+1, y, -z)$, $(-x, y, -z+1)$, $(-x+1, y, -z+1)$. Each of these three positions is related to the original position (x, y, z) by a 2-fold axis. So, in addition to the 2-fold axis along $(0, y, 0)$ there are three more 2-fold axes lying along $(1/2, y, 0)$, $(0, y, 1/2)$, $(1/2, y, 1/2)$ (see Figs 2.37 and 2.38).

In space group P2, the 2-fold axes form 'landmarks' in the plane including **a** and **c**, and x and z are measured from a 2-fold axis chosen as the origin. There are no landmarks along the **b** axis, and the origin of y may be placed at any point along **b**.

The 2-fold axis at the origin creates two asymmetric units (as in Fig. 2.37). As in Fig. 2.27, the operation of the lattice creates three other types of 2-fold axis in the unit cell (Figs 2.37, 2.38).

Space group P2 is the simplest of the monoclinic space groups. Monoclinic space group C2 is described in Box 2.6, as the simplest example of a non-primitive lattice. This

---

Box 2.6 Face centring and screw symmetry: space group C2

Unit cells of the monoclinic lattices, such as P2, have two faces which are rectangular (the faces including the **b** direction). It is possible for one of these to be *face-centred*, so that for every feature near one corner of the cell, there is an equivalent feature near the centre. These features are related by a diagonal translation of half of each of the lattice translations which define a rectangular face (see Fig. 2.34).

You may say, why not define a smaller unit cell by choosing this diagonal translation to be a lattice translation? This is entirely possible. But in consequence, this lattice translation would be neither parallel nor perpendicular to the 2-fold axis direction. We could make a primitive lattice, but at the cost of giving the unit cell a more difficult shape, which would disguise the 2-fold symmetry. For simplicity and convenience, convention chooses a unit cell which has **b** perpendicular to **a** and **c**.

In the symbol C2, the 'C' indicates a lattice centred on the face of the unit cell called C, namely the face whose edges are formed by **a** and **b**. The '2' indicates 2-fold symmetry. Just as in P2 symmetry, the 2-fold axis is placed along **b**, and **a** and **c** are perpendicular to **b**. The origin of coordinates is chosen on a 2-fold axis, but the choice of origin along **b** is arbitrary. The 2-fold symmetry causes any feature A at (x, y, z) to have an equivalent feature B at $(-x, y, -z)$ and there are three more 2-fold axes lying along $(1/2, y, 0)$, $(0, y, 1/2)$, $(1/2, y, 1/2)$ (Fig. 2.39).

The C face centring means that feature A at (x, y, z) generates another feature C at $(x + 1/2, y + 1/2, z)$. Since there are already two equivalent features in the unit cell, generated by the system of 2-fold axes, the face-centring operation generates two more (C and D), and there are four equivalent positions per unit cell. These are at (x, y, z), $(-x, y, -z)$, $(1/2 + x, 1/2 + y, z)$ and $(1/2 - x, 1/2 + y, -z)$.

There is a special relationship between A and D. The x and z coordinates are consistent with a 2-fold axis, not at the origin but at $x = 1/4$, $z = 0$. The y coordinate is, however, shifted from y to $y + 1/2$. The operation relating them is a rotation about a 2-fold axis along **b**, but with a translation of half a unit cell along **b**. This symmetry operation is called a *2-fold screw axis* (symbol 2_1). If the operation is repeated a second time, the total operation is a rotation of 360° (that is, no rotation) and a translation of one unit cell. It is a simple lattice translation. B and C are related by the same operation.

But this is not all! The symmetry of the C2 lattice relates (x, y, z) to $(1/2 - x, y + 1/2, -z)$, so it generates a twofold screw axis at $x = 1/4$, $z = 0$. There are also screw axes at $(1/4, y, 1/2)$, $(3/4, y, 0)$ and $(3/4, y, 0)$ and $(3/4, y, 1/2)$ (Fig. 2.40).

introduces a new type of symmetry operation which combines a rotation with a fractional lattice translation (screw axis of symmetry). Another monoclinic space group, P2$_1$, is introduced in Box 2.7.

Space group P222

In discussing the symmetry of tetrameric molecules, it was shown that if a molecular arrangement possessed two perpendicular 2-fold symmetry axes, a third perpendicular symmetry axis is generated spontaneously (Fig. 2.13). In exactly the same way, when two

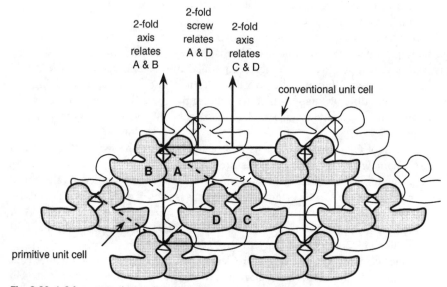

Fig. 2.39 A C-face centred monoclinic structure.

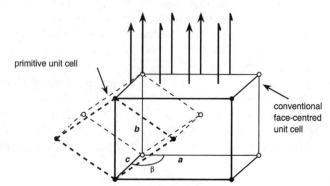

Fig. 2.40 Summary of the symmetry of a C2 lattice. Screw axes are identified by single-barbed arrows.

perpendicular 2-fold axes occur in a lattice, a third is generated. The three perpendicular symmetry axis directions are assigned as lattice dimensions, a, b, c. This requires that the shape of the unit cell shall be a rectangular block, demonstrating the three perpendicular axial directions. Since the coordinates must be exactly perpendicular (or orthogonal) the crystal system is called *orthorhombic*.

The simplest orthorhombic space group is P222 (Box 2.8), where the origin of coordinates x, y, and z is at the intersection of three perpendicular 2-fold axes (Fig. 2.36). Although the axial directions are perpendicular, as in the familiar system of Cartesian coordinates, x, y, and z are still measured as fractions of the unit cell. The 2-fold symmetry axes generate four different orientations for the repeating unit. The asymmetric unit of the crystal is one-quarter of the unit cell.

More space groups!

The space group $P2_1$, already mentioned, is presented in Box 2.9.

There are altogether 65 space groups which use no inversion or mirror operations, which are therefore available to non-centrosymmetric objects like protein molecules. (There are 165 more lattice arrangements which are only possible for centrosymmetric objects. All these were listed independently by Evgraph von Fedorov and Arthur Schoenflies in the nineteenth century—long before any way existed to study the internal structure of crystals.)

By looking in detail at five space groups in this chapter, the most important aspects of internal crystal symmetry have been identified, and will give the reader a working knowledge of lattice symmetry. Not many people remember all the details about the 230 space groups, and the conventions about choice of origin and placement of symmetry elements for each one. Volume A of the *International Tables for Crystallography* has all this information, and a great deal more.

Box 2.7 Work out space group $P2_1$ for yourself!

You may test your insight into these different symmetries, by investigating the space group $P2_1$, the simplest space group with a screw axis. It is very similar to space group P2, but the 2-fold symmetry axis along b at $(0, y, 0)$ is replaced by a 2-fold screw axis. This is an axis which rotates by 180°, but also shifts half a cell along the direction of the axis. The screw axis copies a feature at x, y, z to $-x, y + 1/2, -z$. Find out the positions of other symmetry axes, the coordinates of equivalent positions, and the number of equivalent positions in the unit cell. Box 2.9 (comparable to Box 2.5) and Figs 2.43, 2.44, and 2.45, at the end of the chapter, present space group $P2_1$.

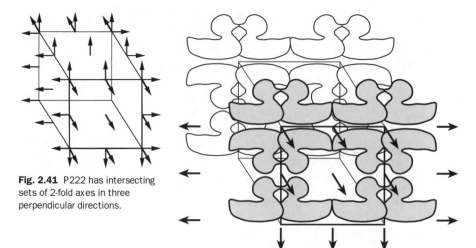

Fig. 2.41 P222 has intersecting sets of 2-fold axes in three perpendicular directions.

Fig. 2.42 Molecular arrangement in space group P222, showing just a few of the 2-fold axes.

Box 2.8 Space group P222

In the symbol P222, the P indicates a primitive lattice, and the 2s indicate three perpendicular 2-fold symmetry axes. In P222, the three axes intersect. The unit cell is defined by the three lattice translations a, b, and c along the directions of these axes, and the positive directions form a right-handed set, as explained in Box 2.1.

The origin of coordinates of x, y, and z is at the intersection of three 2-fold axes. The directions are perpendicular, as in the familiar system of Cartesian coordinates, but x, y, and z are measured as fractions of the unit cell so their units have different lengths.

The 2-fold symmetry along $(0, y, 0)$ causes any feature at (x, y, z) to have an equivalent feature at $(-x, y, -z)$. The lattice symmetry means there are also equivalent positions at $(-x+1, y, -z)$, $(-x, y, -z+1)$, and $(-x+1, y, -z+1)$. Each of these positions is related to the original position (x, y, z) by a 2-fold axis at a different point in the unit cell. So, in addition to the 2-fold axis along $(0, y, 0)$ there are three more 2-fold axes lying along $(1/2, y, 0)$, $(0, y, 1/2)$, $(1/2, y, 1/2)$.

In just the same way, the 2-fold symmetry along $(x, 0, 0)$ causes any feature at (x, y, z) to have an equivalent feature at $(x, -y, -z)$. The feature at $(-x, y, -z)$ generated by the first 2-fold axis produces an equivalent feature at $(-x, -y, z)$. These two 2-fold axes have thus generated four equivalent positions (x, y, z), $(-x, y, -z)$, $(x, -y, -z)$, $(-x, -y, z)$. The first and fourth of these positions are also related by a 2-fold axis along $(0, 0, z)$, and the second and third are related by the same axis. Thus the two 2-fold axes along $(x, 0, 0)$ and $(0, y, 0)$ inevitably generate a third 2-fold axis along $(0, 0, z)$. In the same way as before, 2-fold axes also exist halfway along each cell edge (Fig. 2.41).

There are four equivalent positions in each unit cell. You can think of a 222 tetramer of units, centred on the origin of the unit cell. The whole structure can be built up by allowing lattice symmetry to generate a tetramer at every lattice point (Fig. 2.42).

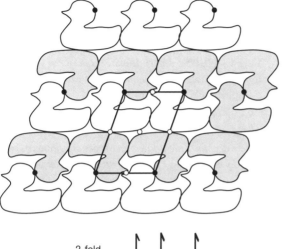

Fig. 2.43 A P2$_1$ structure viewed down the **b** direction. The unshaded molecules are at $y = 0$, and the shaded molecules at $y = 1/2$. There are 2-fold screw axes at the corners of the unit cell, and also at positions indicated by white circles.

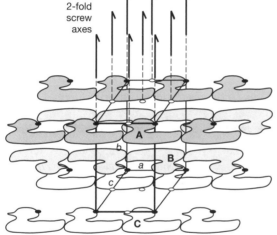

Fig. 2.44 Arrangement of units in a P2$_1$ lattice. Units facing one way are at the top and bottom of the cell, those facing the other are halfway in between. Objects A, B, and C are related by a 2-fold screw operation.

Box 2.9 Space group P2$_1$

In the symbol P2$_1$, the P indicates a primitive lattice, and the 2$_1$ indicates a 2-fold screw axis. The 2-fold screw axis is conventionally placed at $(0, y, 0)$.

The origin of coordinates is chosen on a 2-fold screw axis, but the choice of origin along **b** is arbitrary. The 2-fold screw symmetry causes any feature at (x, y, z) to have an equivalent feature at $(-x, y + 1/2, -z)$. The lattice symmetry means there are also equivalent positions at $(-x + 1, y + 1/2, -z)$, $(-x, y + 1/2, -z + 1)$, and $(-x + 1, y + 1/2, -z + 1)$.

Each of these positions is related to the original position (x, y, z) by a 2-fold screw axis. So, in addition to the 2-fold screw axis along $(0, y, 0)$ there are three more 2-fold screw axes lying along $(1/2, y, 0)$, $(0, y, 1/2)$, $(1/2, y, 1/2)$. This is viewed along the **b** axis in Fig. 2.43, and in another direction in Fig. 2.44.

There are two equivalent positions in each unit cell. The symmetry is summarized in Fig. 2.45.

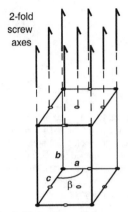

2-fold screw axes

Fig. 2.45 Summary of the symmetry of a P2₁ lattice.

Further reading

The definitive reference book concerning crystal symmetry is:
Hahn, T. (ed.) (1995). *International tables for crystallography. Volume A. Space-group symmetry.* International Union of Crystallography/Kluwer, Dordrecht.

Most of the crystallography textbooks listed after Chapter 1 present a basic introduction to crystal symmetry. They give a more complete account, including other types of symmetry operation which do not apply to chiral structures, for example:
Glusker, J.P. (1994). *Crystal structure analysis for chemists and biologists.* VCH, New York.

A clear and full account of crystallographic symmetry can be found in:
Hammond, C. (1997). *The basics of crystallography and diffraction,* Chapters 2, 3, and 4. International Union of Crystallography/Oxford University Press, Oxford.

For more general reading about symmetry, a classic is:
Weyl, H. (1952). *Symmetry.* Princeton University Press, New Jersey.

Symmetry in biology is the theme of:
Thompson, D.W. (1917). *On growth and form.* Cambridge University Press, Cambridge; *republished in a shorter edition as:*
Thompson, D.W. (1961). *On growth and form* (ed. J.T. Bonner). Cambridge University Press, Cambridge.

The artist and engraver Maurits Escher exploited symmetry and pseudo-symmetry with wonderful results. Many collections of his work are available. Specially relevant is
McGillavray, C.K. (ed.) (1976). *Symmetry aspects of M.C. Escher's periodic drawings.* International Union of Crystallography/ Bohn, Schletema and Holkema, Utrecht.

3

Waves

In this very short chapter, some basic facts about waves are presented. Attached to this short chapter are several boxes which give the fundamental mathematical basis for understanding waves in a more quantitative fashion.

There are many physical examples of waves. Waves in water are perhaps the most familiar example. A water wave is created by a disturbance in the height of the water surface. The amount by which the height of the water is disturbed by the wave is called its amplitude.

Another important form of wave is sound, which is a variation of pressure in a gas or liquid (or of stress in a solid). But for our purposes the most important waves are electromagnetic waves, specifically X-rays, with wavelengths of an Ångström or so. Electromagnetic waves create a disturbance in both the electric field and the magnetic field: usually the wave is represented by its electric component.

All these waves carry energy. The rate of energy transfer is called the intensity, and at a given wavelength the intensity is proportional to the square of the amplitude. Detectors of X-rays, discussed in Chapter 1, respond to the quantity of energy delivered by the beam, which is also proportional to the number of photons. The amplitude of any wave is thus proportional to the square root of its intensity.

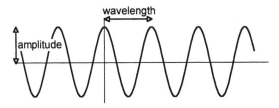

Fig. 3.1 A sinusoidal wave, showing wavelength and amplitude.

Sinusoidal waves

The most simple form of wave is a sinusoidal disturbance which moves forward at a fixed velocity. 'Sinusoidal' means shaped like a sine wave (Fig. 3.1). For reasons that will emerge, we will work more often with a cosine function, which is just the same shape as a sine wave, but which has its origin at a maximum point of the wave.

A sinusoidal wave can be described by several properties:

- the *wavelength*, which is the distance from one peak to the next;
- the *amplitude*, which is the height of the wave peak above its mean level;

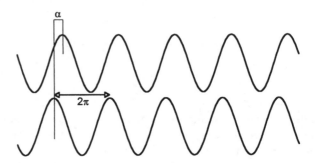

Fig. 3.2 Two waves are shown, differing in phase by α. A phase change of 2π would shift one peak of the wave to the next.

Box 3.1 Waves

A pure cosine wave can be expressed as:

$$W = A \cos[(2\pi x/\lambda) + \phi]. \tag{1}$$

Here W is the quantity which the wave disturbs (e.g. height above the mean level); x is the direction of propagation; A is its amplitude, λ its wavelength, and ϕ is its phase. When x changes by λ, the bracket changes by 2π (2π radians $= 360°$), and the cosine function has an identical value. Thus, the wave exactly repeats itself over each wavelength (Fig. 3.1).

This might be the form of the wave at some particular time, say $t = 0$. To make the wave move in time, a phase factor is needed which depends upon the time t. If the wave moves with velocity c, the peak of the wave moves forward by a distance ct. Eqn 1 is changed to:

$$W = A \cos[(2\pi(x - ct)/\lambda) + \phi]. \tag{2}$$

If $t = 0$, this is exactly the same as eqn 1. One peak of the wave is at the point where the argument of the cosine function [in brackets] is exactly 0. If the time changes to t, this position is changed by ct. As t increases, x at the peak has to increase to keep the value of $(x - ct)$ unchanged. Notice that the negative sign causes the wave to move forward (along $+x$) as the time advances.

- the *phase*, which specifies where the peak of the wave is, relative to an origin of measurement at the position $x = 0$, and the time $t = 0$;
- the *wave velocity*, which is the velocity at which the wave advances along the propagation direction.

Mathematical formulae for expressing a wave are given in Box 3.1. The wave in Fig. 3.1 is a cosine wave with phase $\phi = 0$, and has a peak at the origin. A change of phase causes a displacement of the wave along its axis of propagation (Fig. 3.2); if ϕ is changed by the angle 2π (or $360°$), there is a displacement of a whole wavelength, which brings the wave back to look exactly as it started. A displacement by a whole number wavelengths is therefore a symmetry operation for a cosine wave.

A travelling wave also repeats in time. For our purposes we can usually think of the wave as a snapshot taken at a fixed time, but the generalization to other times, showing the wave moving forward along the x axis, is quite simple (Box 3.1).

After the time it takes the wave to advance by one wavelength, it looks exactly as it started. The *frequency* of the wave, which gives the number of vibrations per second, is the reciprocal of this length of time.

Complex numbers

To understand the material in some of the later boxes, it will be necessary to have some familiarity with complex numbers. A brief introduction to complex numbers, including only what you need to know to understand the later boxes, is given in Boxes 3.2, 3.3, and 3.4.

An ordinary number, also known as a *real* number, only represents one quantity. We may consider a complex number to be a way of representing two quantities by a single entity, which has two components. It can represent the amplitude and phase of a wave. Convenient methods of analysing wave properties rely on the use of complex numbers.

A complex number can be expressed graphically as a point on a two-dimensional diagram (Fig. 3.3), and can often be thought of as a vector from the origin to this point. The complex number is composed of a *real part*, drawn on the horizontal or x axis, and an *imaginary part* shown as a vertical displacement from the x axis. To show it is an imaginary part, the vertical displacement is written iy. The symbol i represents an operation which turns a real number y into an imaginary number iy. Notice that in Fig. 3.3 (a type of diagram displaying a complex number, known as an *Argand diagram*) the operation i results in rotating a real number (which would be plotted on the horizontal axis) into an imaginary number plotted vertically. On the diagram, the operation i rotates the vector representing the number by $\pi/2$ ($90°$) anticlockwise.

A complex number z can also be defined by the length of the vector which represents it, and the direction of this vector. The length of the vector is its *magnitude*, a real number written as $|z|$, and the angle ϕ it makes with the real (or horizontal) axis will be called its *phase* in this text. (Mathematicians use the technical word *argument* for this angle.)

Addition of two complex numbers (Box 3.3) results in adding the two corresponding vectors (Fig. 3.4). Multiplication of two complex numbers includes a rotational operation (Box 3.3), illustrated in Fig. 3.5.

Figure 3.6 presents another useful relation between complex numbers, and shows what is meant by the *complex conjugate* of a complex number (Box 3.4). Multiplying a complex number by its conjugate gives a real number equal to the square of its magnitude.

Box 3.2 Complex numbers

For present purposes, a complex number is a way of representing a point in a two-dimensional space by a single number. It may also be used to represent the amplitude and phase of a wave by a single quantity.

A complex number z may be written as $x+iy$, where x is called the '*real part*' of the number and y is called its '*imaginary part*'. These can simply be imagined as horizontal and vertical coordinates which represent a point z on a two-dimensional sheet of graph paper. 'Real' and 'imaginary' are just convenient names for the two parts of the complex number.

A complex number may also be written as $|z|e^{i\phi}$. The *amplitude* $|z|$ is the magnitude of the complex number z. It is the distance of the point z from the origin. ϕ is its *phase*, that is the angle between the real axis (horizontal axis) and the line joining the origin to the position z. There is no need to worry about the meanings of the individual symbols in $e^{i\phi}$. Just think of it as an operation which takes the real number $|z|$ and turns it through an angle ϕ. The two ways of writing the complex number z as $|z|e^{i\phi}$ or as $x+iy$ imply that

$$|z|e^{i\phi}=x+iy. \tag{1}$$

As shown in Fig. 3.3, this means that:

$$z=x+iy=|z|(\cos\phi+i\sin\phi)=|z|e^{i\phi}, \tag{2}$$

so

$$x=|z|\cos\phi \quad \text{and} \quad y=|z|\sin\phi.$$

Also, since the x and y axes are perpendicular,

$$|z|^2=(x^2+y^2). \tag{3}$$

The operator i is the square root of minus one!

So far, the symbol i has just been used as a prefix which identifies the imaginary part of a number. It is best to think of i as an operation which can be carried out on a number. Multiplying a real number y by i carries out an operation to make it into an imaginary number iy. On the *Argand diagram*, which is the name of a diagram drawn in the complex plane, like Fig 3.3, the same operation rotates the vector by $\pi/2$ (a right angle) anticlockwise. If the imaginary number iy is operated again by i to give i^2y, it is rotated anticlockwise by a further $\pi/2$ angle to become a real number of the opposite sign $-y$. Thus, carrying out the operation i twice on a number changes its sign:

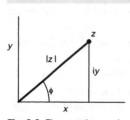

Fig. 3.3 The complex number z is composed of a real part x and an imaginary part y. The complex number may also be defined by its magnitude $|z|$ and its phase angle ϕ.

$$i(iy)=i^2y=-y, \quad \text{or} \quad i^2=-1. \tag{4}$$

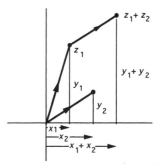

Fig. 3.4 Addition of two complex numbers is like adding two two-dimensional vectors.

Box 3.3 Algebra rules for complex numbers

Adding complex numbers

Adding two complex numbers is just like adding two vectors on the Argand diagram (Fig. 3.4). The real parts and the imaginary parts may be added independently, giving a complex number whose real and imaginary components are each the sum of the two starting components:

$$z_1 + z_2 = (x_1 + iy_1) + (x_2 + iy_2)$$

$$= (x_1 + x_2) + i(y_1 + y_2). \tag{1}$$

Multiplying two complex numbers

Notice that eqn 2 of Box 3.2 shows that

$$e^{i\phi} = \cos\phi + i\sin\phi \tag{2}$$

As explained in Box 3.2, $e^{i\phi}$ represents the operation of turning anticlockwise through ϕ. If the angle ϕ is $\pi/2$, this operation is the same as operating by i; if the angle ϕ is π, the operation is the same as multiplying by -1. Thus

$$ze^{i\pi/2} = iz; \quad \text{and} \quad ze^{i\pi} = -z. \tag{3}$$

With these extra algebraic rules given in eqn 3, all the normal rules of algebra apply. Just as $2^2 \times 2^3 = 2^{2+3} = 2^5$ (or $4 \times 8 = 32$), so

$$|z| e^{i\phi} \times e^{i\psi} = |z| e^{i(\phi+\psi)}. \tag{4}$$

This can be read as 'The operation of rotating through ϕ followed by rotation through ψ is the same as rotating through $(\phi+\psi)$' (Fig. 3.5). Remember that $|z| e^{i\phi}$ may be written as a complex number $x+iy$ (as in Box 3.2, eqn 1). Also $e^{i\psi}$ may be written as $\cos\psi + i\sin\psi$ (eqn 2). Making these changes in eqn 4,

$$(x+iy)e^{i\psi} = (x+iy)(\cos\psi + i\sin\psi)$$
$$= (x\cos\psi - y\sin\psi)$$
$$+ i(x\sin\psi + y\cos\psi), \tag{5}$$

using $i^2 = -1$.

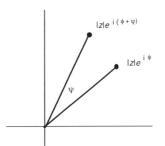

Fig. 3.5 Multiplication by a complex number of magnitude 1, but phase angle ψ, is like rotating through the angle ψ. This complex number may be written as $e^{i\psi}$.

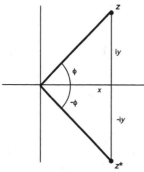

Fig. 3.6 A complex number and its conjugate. If the real part of the number is x and its imaginary part is iy, the complex conjugate has the same real part x but its imaginary part is $-iy$.

Box 3.4 The complex conjugate

If $z = x + iy$, the quantity $z = x - iy$ is called its complex conjugate, written $z^\star$. Similarly, the complex conjugate of $|z|e^{i\phi}$ is $|z|e^{-i\phi}$ (Fig. 3.6). As shown in the figure, the complex conjugate appears on an Argand diagram as the 'mirror image', where the real axis forms the mirror. A special property of complex conjugate pairs is that their product is always real, and is the square of the amplitude:

$$z.z^\star = (x+iy) \cdot (x-iy) = x^2 + y^2 = |z|^2 \qquad (1)$$

or

$$|z|e^{i\phi} \cdot |z|e^{-i\phi} = |z|^2 (e^{i\phi} \cdot e^{-i\phi}) = |z|^2. \qquad (2)$$

Box 3.5 Waves represented by complex numbers

It is often mathematically convenient to use the complex number notation to represent a wave. The quantity W which forms the wave is a real quantity, and the wave at $t = 0$, for example, could be written:

$$W = \mathrm{Re}\{A \exp[i(2\pi x/\lambda + \phi)]\}. \qquad (1)$$

There are two new points of notation in this equation. The symbols Re{ } mean 'the real part of the complex number within the braces'. The symbol exp[z] is used for typographical convenience to replace e^z. This allows complicated exponents to be written out more clearly, in a larger typeface. Since

$$\exp[i\phi] = \cos\phi + i\sin\phi$$

(see Box 3.3, eqn 2), the real part of exp[$i\phi$] is just $\cos\phi$, and eqn 1 above is a restatement of eqn 1, Box 3.1. Eqn 2 of Box 3.1 could be rewritten in exactly the same way.

The merit of this notation is that it often simplifies complex trigonometric expressions which all contribute to the phase of the wave. The phase is just the angle ϕ on the Argand diagram, which represents the complex exponent $i\phi$.

In practice, the Re{ } notation is often omitted. If a number which is bound to be real is found equated to a complex number, you may assume that the Re{ } notation is implied.

Given its wavelength, a sinusoidal waveform can be completely characterized by a single complex number which specifies its amplitude and phase (Box 3.5). The intensity of the wave is proportional to the square of its amplitude, that is the square of the magnitude of this complex number, which is also the product of this complex number with its conjugate.

Fourier analysis

So far, only a simple sinusoidal wave has been discussed. In reality, many other shapes of waves are encountered, which may still be exactly repetitive. A sinusoidal sound wave produces a 'pure' note, which sounds like a tuning fork, or like the 'hum' from a transformer on the alternating mains supply. Different kinds of musical instrument create sound waves of different shapes, producing different sound qualities such as the roundness of a trumpet note or the sharpness of a bowed violin. While the performer holds the note steady, it is ideally an exact waveform, repeating itself precisely at a certain frequency, known as the 'fundamental' frequency (Fig. 3.7).

Joseph Fourier proposed a way of analysing complicated repetitive waves (such as Fig. 3.7) by representing them as the sum of a series of waves whose frequencies are integral multiples of a fundamental frequency (Box 3.6). In a musical note, these components are called the 'harmonics' of the fundamental sound frequency, and control the quality of the sound which is heard. The flute note, for example, is produced almost entirely by the fundamental frequency and a sound of double the frequency (Fig. 3.8). The oboe and the violin have sharper tones because many higher harmonics are included. A strong seventh harmonic is characteristic of the sound of a violin.

Fourier analysis of crystals

Fourier's methods can be applied to almost any repetitive function to analyse the harmonics that compose it. In crystallography, the most important application of Fourier's methods is in the analysis of the repetitive structures in crystals. These repetitive structures can be analysed into harmonics in exactly the same way as the oboe note. But since we are usually interested in three-dimensional structure, the analysis has to be carried out in three-dimensional space. More will be said about the generalization to three dimensions in Chapter 5. The general principles can be presented perfectly well in one dimension (Boxes 3.6, 3.7, and 3.8).

Using these methods, the variation of electron density in a crystal can be used to determine the relative amplitude of each of its Fourier components, and also their phases, by a *direct* Fourier transformation. Alternatively, proceeding in the opposite way, the amplitudes and phases of the different components can be used to calculate the density ρ of electrons in the crystal. This is called an *inverse* Fourier transformation. The two types of operation are very closely linked. Box 3.9 describes the similarities and the differences between them.

The process of building up an image of an atomic structure from its Fourier components is illustrated in detail in the next chapter.

Box 3.6 Fourier series

The representation of a complicated waveform repeating along the x-axis, $W(x)$, by the sum of a series of cosine waves is called a Fourier series. We shall define the unit of length so the waveform repeats exactly in each unit. The Fourier series representing such a wave is:

$$W(x) = A_0 + A_1 \exp[i(2\pi x + \phi_1)] + A_2 \exp[i(2 \cdot 2\pi x + \phi_2)]$$
$$+ A_3 \exp[i(3 \cdot 2\pi x + \phi_3)] + \cdots \tag{1}$$

This shows the complex waveform expressed as the sum of a series of waves with progressively shorter wavelengths $\lambda = 1, 1/2, 1/3 \ldots$ each with its own amplitude A_n and phase ϕ_n. Note that an A_0 term is included, which sets the mean level of W. This term has infinite wavelength, and does not need a phase factor.

The previous equation can be written more compactly as:

$$W(x) = \sum_{n=0}^{\infty} A_n \exp[i(2\pi n x + \phi_n)]. \tag{2}$$

The complex numbers $A_n \exp[i\phi_n]$ are called the *Fourier coefficients* of the waveform $W(x)$. A_n and ϕ_n are the *amplitude* and *phase* of the Fourier coefficient. The Fourier coefficient $A_n \exp[i\phi_n]$ can be written as the complex number F_n. The waveform can then be written:

$$W(x) = \sum_{n=0}^{\infty} F_n \exp[2\pi i n x]. \tag{3}$$

In crystallography, we are dealing with structures which repeat in every unit cell of the crystal. The scattering density is written as $\rho(x)$. It can be a complicated waveform and it repeats in every unit cell. The harmonic n is usually identified by h in crystallography. To fit in with other conventions, we will replace n in eqns 1–3 with $-h$. In crystallography, F_h is called the *structure factor*, which may be written as $|F_h| \exp[i\phi_h]$, where $|F_h|$ and ϕ_h are the amplitude and phase of the scattered wave numbered h. $|F_h|$ is called the *structure amplitude*. Introducing this notation, eqn 3 becomes:

$$\rho(x) = \sum_{h=0}^{\infty} |F_h| \exp[i\phi_h] \, \exp[-2\pi i h x]$$
$$= \sum_{h=0}^{\infty} F_h \exp[-2\pi i h x]. \tag{4}$$

This compact expression defines a repetitive scattering density in terms of its structure factors.

In practice, we need to express the electron density as a function varying in a three-dimensional space. The generalization from the one dimension used here to three dimensions will be shown in Chapter 5.

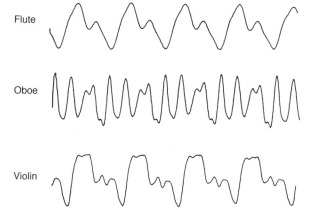

Flute

Oboe

Violin

Fig. 3.7 Waveforms produced by various musical instruments. (Reproduced from Johnston (1989) by permission of the Institute of Physics.)

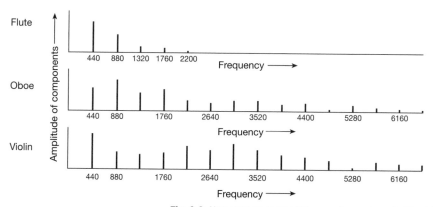

Fig. 3.8 Harmonic analysis of the waveforms shown in Fig. 3.7, where the fundamental frequency is 440 cycles/s. (Reproduced from Johnston (1989) by permission of the Institute of Physics.)

Box 3.7 The orthogonality rule: a result used in Fourier analysis

Demonstrations using Fourier transforms often use a special property of sine and cosine functions. It concerns the total area under the curve generated by a product of two sinusoidal waves, that is the integral of the product, taken over one whole repetition. The area can be written, in the customary notation:

$$\int_0^1 \exp[2\pi ihx] \cdot \exp[2\pi ikx]\,dx,$$

h and k being integers.

When $k=-h$, this product is $\int_0^1 \exp[0]\,dx=1$. But in all other cases, its value is precisely zero. This is because the product $\exp[2\pi ihx] \cdot \exp[2\pi ikx]$ represents a waveform which is positive exactly as often as it is negative, during one whole repetition, so that the integral is precisely zero (Fig. 3.9).

This means that if a series of such products are added, many of the terms are precisely zero. Consider a summation over all values of k, for example,

$$\int_0^1 \left(\sum_k \exp[2\pi ihx] \cdot \exp[2\pi ikx] \right) dx.$$

All the members of the summation contribute precisely zero to the integral, except that for which $k=-h$. This non-zero term is $\exp[0]$, which is 1, and therefore the result of the whole integration is 1.

This type of result is frequently used in analysing a double summation such as

$$\int_0^1 \sum_h \sum_k (Z(h, k) \exp[2\pi ihx] \cdot \exp[2\pi ikx])\,dx,$$

where $Z(h, k)$ represents some complex number which depends on both h and k. For every term in the summation over h, only one term in the summation over k is non-zero. This is the term where $k=-h$. Following the same argument as before, the integral becomes

$$\int_0^1 \sum_h (Z(h, -h) \exp[2\pi ihx])\,dx.$$

The result

$$\int_0^1 \exp[2\pi ihx] \cdot \exp[2\pi ikx]\,dx=1 \quad \text{if } k=-h,$$
$$= 0 \text{ otherwise}, \qquad (1)$$

is said by mathematicians to express the *orthogonality* of the trigonometric functions represented by $\exp[2\pi ihx]$. When it is used here, it will be referred to as the *orthogonality rule*.

Box 3.8 How to determine the Fourier coefficients of a waveform

Given the waveform $\rho(x)$, how can we determine its Fourier coefficients? Here is the recipe to determine one particular Fourier coefficient identified as h.

What we do is to multiply $\rho(x)$ by a cosine wave which repeats h times in one repeat length of ρ, and integrate this product over the repeat length. This picks out just the contribution to $\rho(x)$ which has wavelength $1/h$, and provides the hth Fourier coefficient F_h. The orthogonality rule shows that all the other contributions cancel themselves out.

In mathematical symbols, if the electron density ρ is weighted by $\exp[2\pi ihx]$ and expanded as a Fourier series as in Box 3.6, eqn 4, it gives

$$\int_0^1 \rho(x)\, \exp[2\pi ihx]\, dx$$

$$= \int_0^1 \left\{ \sum_k F_k \exp[-2\pi ikx] \right\} \exp[2\pi ihx]\, dx.$$

Note that the symbol k is used to identify the Fourier coefficients of ρ because h is already in use. The orthogonality rule (eqn 1, Box 3.7) shows that in the integration, the terms cancel themselves out unless $h=k$. When $h=k$, the integral over the exponentials is exactly 1; otherwise it is zero. Only the hth term in the summation makes any contribution. The Fourier component F_h is constant in the integration, not dependent on x, so the integral may finally be written:

$$\int_0^1 \rho(x)\, \exp[2\pi ihx]\, dx = F_h. \tag{1}$$

This equation states mathematically the verbal statement at the beginning of this box, that the hth Fourier coefficient may be calculated by multiplying the repeating function by $\exp[2\pi ihx]$, and integrating the result over one repetition of the waveform.

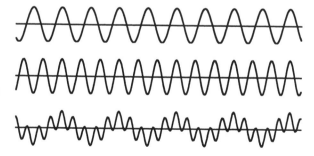

Fig. 3.9 The waveforms $y=\cos(2\pi \cdot 2x)$ and $y=\cos(2\pi \cdot 3x)$, and their product. The product of two cosine terms is positive as often as it is negative, so that the integral over a whole repeat is zero.

Box 3.9 Fourier series and Fourier integrals

Direct and inverse Fourier transformations

Compare eqn 4 of Box 3.6

$$\rho(x) = \sum_{h=0}^{\infty} F_h \exp[-2\pi ihx]$$

and eqn 1 of Box 3.8

$$F_h = \int_0^1 \rho(x) \exp[2\pi ihx] \, dx.$$

They have some differences and some similarities. In Box 3.6 a sum is taken over all the Fourier coefficients of the waveform. It is called a Fourier series. The wave is formed by adding all the Fourier coefficients, each weighted by $\exp[-2\pi ihx]$ which expresses the waveform representing this term in the Fourier series. In Box 3.8, an integral is taken over one period of the whole waveform, and is called a Fourier integral. Each point in the waveform is weighted by $\exp[2\pi ihx]$. Both these processes, going from the waveform to the Fourier coefficients, or from the Fourier coefficients to the waveform, are called *Fourier transformation*. Box 3.8 presents a *Fourier integral* in which the waveform is weighted by $\exp[2\pi ihx]$ to calculate its Fourier coefficients. The summation in Box 3.6 is a *Fourier summation*. It uses the weighting factor $\exp[-2\pi ihx]$. Because of the negative sign, it is called the *inverse Fourier transform* of the Fourier coefficients, which may be used to generate the wave.

Further reading

The material in this chapter is a small part of the material on complex numbers and Fourier transforms that can be found in any number of undergraduate textbooks.

Clear and detailed sections on complex numbers and Fourier analysis can be found in:
Steiner, E. (1996). *The chemistry maths book.* Oxford University Press, Oxford.

A more extensive chapter on complex numbers, leading into Fourier series, and another chapter on Fourier analysis, are in
Lyons, L. (1998). *All you wanted to know about mathematics but were afraid to ask: mathematics for science students.* Cambridge University Press, Cambridge.

A readable classic book on mathematics, which can be dipped into to get outlines of topics such as complex numbers, is
Courant, R. and Robbins, H. (1996). *What is mathematics?* (2nd revised edn) (revised by I. Stuart). Oxford University Press, Oxford.

4

Diffraction

Diffraction refers to the effects observed when light is scattered into directions other than the original direction of the light, without change of wavelength.

An X-ray photon may interact with an electron and set the electron oscillating with the X-ray frequency. The oscillating electron may radiate an X-ray photon of the same wavelength, in a random direction, when it returns to its unexcited state. Other processes may also occur, akin to fluorescence, which emit X-rays of longer wavelengths, but these processes do not give diffraction effects.

Just as we see a red card because red light is scattered off the card into our eyes, objects are observed with X-rays because an illuminating X-ray beam is scattered into the X-ray detector. Our eye can analyse details of the card because its lens forms an image on the retina. Since no X-ray lens is available, the scattered X-ray beam cannot be converted directly into an image. Indirect computational procedures have to be used instead.

X-rays are penetrating radiation, and can be scattered from electrons throughout the whole scattering object, while light only shows the external shape of an opaque object like a red card. This allows X-rays to provide a truly three-dimensional image. When X-rays pass near an atom, only a tiny fraction of them is scattered: most of the X-rays pass further into the object, and usually most of them come straight out the other side of the whole object. In forming an image, these 'straight through' X-rays tell us nothing about the structure, and they are usually captured by a beam stop and ignored.

This chapter begins by explaining that the diffraction of light or X-rays can provide a precise physical realization of Fourier's method of analysing a regularly repeating function. This method may be used to study regularly repeating distributions of scattering material. Beginning in one dimension, examples will be used to bring out some fundamental features of diffraction analysis. Graphic examples of two-dimensional diffraction provide further demonstrations.

Although the analysis in three dimensions depends on exactly the same principles, diffraction by a three-dimensional crystal raises additional complications. Crystal diffraction will be presented in Chapter 5.

Fraunhofer diffraction: an example of a Fourier analysis

We are going to begin by thinking about a regular array of scattering objects, and we shall start off in one dimension, with a regular array of scattering particles along a line. The same effects arise in two dimensions when light from a distant source is scattered through the regular mesh of a handkerchief. Scattering by the regular three-dimensional lattice of a crystal is the most important application of these ideas.

For Fourier's analysis to apply to diffraction, the diffracting sample must be far away from the source (compared to its size) so that the X-rays from a point in the source are effectively parallel, and also the detector distance must be large compared to the sample size. These conditions are well satisfied in the usual types of X-ray diffraction experiment. Diffraction under these conditions is known as *Fraunhofer diffraction* (Fig. 4.1).

In the most usual case, X-ray diffraction uses an X-ray beam having all component waves of the same wavelength. Such an X-ray beam is called *monochromatic*. This chapter concentrates on monochromatic diffraction. The less usual case, where many wavelengths are used simultaneously (known as *Laue diffraction* for X-rays), is mentioned later.

Suppose a parallel beam of monochromatic light is projected onto a line perpendicular to it, which has many scattering particles regularly spaced along it (Fig. 4.2). We are going to assume that the spacing of the particles is bigger than the wavelength of the light. Each particle scatters light in all directions, but the waves scattered from the different particles tend to cancel each other out, because they have different phases.

There is, however a direction where the light scattered from two adjacent particles reaches the same wavefront by paths which differ by exactly one wavelength. For a wavefront travelling in this direction, the light scattered by all the particles is exactly in phase, because the light paths differ by exactly a whole number of wavelengths.

Box 4.1 shows how to work out the direction of this scattered light beam. It also shows that there are other directions which have all the particles scattering in phase. In these directions the light paths differ by two wavelengths, or three wavelengths, or by another whole number of wavelengths (Fig. 4.1). The repetitive scattering object acts as a *diffraction grating*, and the light beams scattered in the particular directions are called

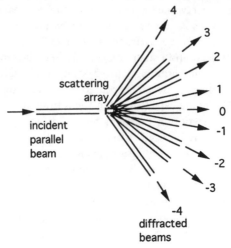

Fig. 4.1 Fraunhofer diffraction of a parallel beam by a regular array of scatterers.

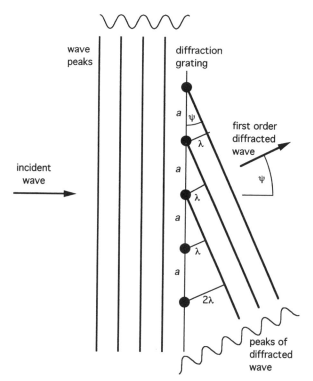

Fig. 4.2 First-order diffraction by a series of equally spaced scattering points.

Box 4.1 A simple diffraction grating

A parallel beam of light, wavelength λ, is projected on to a line of scattering matter perpendicular to it, with particles regularly spaced along this line at a spacing a. Each scattering particle scatters light equally in all directions (Fig. 4.1).

Consider how much light is scattered in a direction making an angle ψ with the scattering line. First, notice that if $\sin \psi = \lambda/a$, the light scattered in this direction from two adjacent particles is exactly in phase, because one scattered beam has to travel exactly one wavelength further than the next. Notice also, that for a pair of scattering particles ha apart (h being an integer) the light scattered in this direction from these two particles is again exactly in phase, because one light beam has to travel exactly h wavelengths further (Figs 4.2, 4.3). For two adjacent units, the path difference for the scattered beam, $a \sin \psi$ equals $h\lambda$. Therefore, light may be scattered strongly in all the directions given by $\sin \psi = h\lambda/a$, because scattering from all the particles is exactly in phase.

The light scattered in a direction where $\sin \psi$ is nearly $h\lambda/a$, but not quite, will be scattered with a slightly different phase from every particle. If we go far enough along the line of particles, we shall come to one that scatters out of phase with the first particle, cancelling it out. Provided the line of scattering particles is long enough, all the scattered light is cancelled out, unless it is exactly in one of the directions ψ_h, where $\sin \psi_h = h\lambda/a$. For an infinite array, a significant amount of energy is scattered only in these directions.

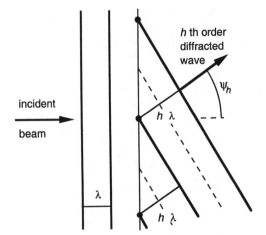

Fig. 4.3 Second-order diffraction by a series of equally spaced scattering points. In this example the path lengths differ by two wavelengths and the order *h* is 2.

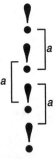

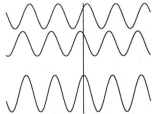

Fig. 4.4 Regular repeating array of scattering density. The density at any point is the same as the density at a distance *a* from it.

Fig. 4.5 The sum of two waves of the same wavelength, which have different phases. The wave representing the sum has the same wavelength as the original waves, but the amplitude and phase are different.

orders of diffraction. The light scattered nearest to the direction of the unscattered ray is called the first order, because there is one wavelength difference in the distance travelled by light from adjacent scatterers (Fig. 4.2). Other directions of scattering are identified by an integer, *h*, which increases as the scattering angle increases (Fig. 4.3). Part of the energy of the incident wave will be scattered into these various directions.

The diffraction grating does not have to have only one kind of regularly spaced particles. It may have any repetitive distribution of scattering matter along the line. So long as it repeats exactly, each particular bit of scattering matter has identical bits at other points along the grating, separated by the repeat distance of the grating. All these bits of scattering matter scatter in phase, in the direction of an order of diffraction (Fig. 4.4). The total scattered beam is the total of all these scattering components. A diffraction grating with varying scattering density in the repeat unit still scatters light only in the directions identified with orders of diffraction.

Consider one particular scattered beam, of order *h*. The contributions to the scattered beam from different parts of the repeating distribution will have different phases. Because of these phase differences, the wave comprising the total scattered beam is not just a wave having an amplitude equal to the sum of all the component scattering amplitudes. We say the waves *interfere* with one another (Fig. 4.5 and Box 4.2). The total scattered waves will also have a different phase compared to the incident light beam.

When we work out the amplitudes and phases of the waves scattered in different directions by this kind of diffraction grating (Box 4.3) something useful emerges. The amplitudes and phases of the scattered waves represent precisely the Fourier transform (described in Chapter 3) of the scattering density. The diffraction process

---

Box 4.2 A repeating one-dimensional scattering object

Following the convention adopted in crystallography, the unit of length will be taken as the repetition distance along the line of the lattice. According to this definition, $a = 1$ unit. The coordinate x used to measure distance in this direction will therefore measure it as a fraction of the repetition distance.

Now suppose that there is scattering matter all along the line, and that the density of scattering matter, $\rho(x)$, repeats exactly in each unit (Fig. 4.4). In proceeding from one point to an equivalent one, x increases by one unit. Thus $\rho(x) = \rho(x+1) = \rho(x+n)$, where n is any integer. In the directions ψ_h all these equivalent points scatter in phase. This applies for every point, whatever the value of x. Exactly as with the simple grating, it can scatter in the directions ψ given by $\sin \psi = h\lambda/a$, but in no others.

Box 4.3 Diffraction creates a Fourier transform

The contribution to the scattering in direction ψ_h from a very short length of scattering matter dx, so short that $\rho(x)$ can be considered constant within it, will be proportional to $\rho(x)dx$. But the scattered light will have its phase shifted relative to the phase of scattering from $x = 0$. For the hth order of diffraction, there would be a phase shift of $2\pi h$ as x goes from 0 to 1. So the phase shift at point x would be $2\pi hx$.

By adding up the scattering from all the lengths dx, as x traverses a unit from 0 to 1, we can work out the total wave scattered by one whole scattering unit in the direction ψ_h. The scattering by this unit into the hth order of diffraction is formed by adding together all these contributions to form the integral $\int_0^1 \rho(x) \exp[2\pi ihx] \, dx$. This quantity is called the *structure factor* for this order of diffraction, and is a complex number with an amplitude $A = |F_h|$ and phase α_h:

$$F_h = |F_h| \exp[i\alpha_h] = \int_0^1 \rho(x) \exp[2\pi ihx] \, dx. \tag{1}$$

This equation corresponds exactly to Box 3.8, eqn 1.

Similarly $\rho(x)$ may be derived from an inverse Fourier summation over all the scattered rays as in Box 3.6, eqn 4.

$$\rho(x) = \sum_h F_h \exp[-2\pi ihx] = \sum_h |F_h| \exp[-2\pi ihx + \alpha_h]. \tag{2}$$

The structure factor F_h is a complex number representing the amplitude and phase of the hth scattered wave.

represents a kind of analogue computer for working out the Fourier transform of the scattering density function.

In these examples, the scattering function is one-dimensional. We will stay with one-dimensional scattering functions for a while, to bring out some general points in the simplest way.

X-ray and neutron detectors only observe the intensity

As discussed already in Chapter 1, detectors of X-rays measure the energy of light entering them. They are insensitive to the phase. The energy of light in a diffracted beam is proportional to the square of its amplitude: all we can detect is the intensity for each diffracted wave (proportional to the square of the amplitude). This applies to all available detectors of short wavelength electromagnetic waves, and to detectors of matter waves.

The diffraction grating is a device for producing waves which represent the Fourier transform of the scattering density. But if we want to work backwards to find the scattering density function $\rho(x)$ from the diffracted rays, we need to know the phases of the scattered waves as well as their amplitudes (see Box 4.3, eqn 2). Since these phases cannot be observed directly, the difficulty in determining the phase of a scattered wave is referred to as the *phase problem*. Ways of solving this problem are presented in Chapters 7, 8, and 9.

Centrosymmetry and phase angles

A repeating centrosymmetric distribution (such as the one shown in Fig. 4.6) can always be built up from cosine waves whose maximum or minimum comes at a centre of symmetry. The cosine waveform is centrosymmetric, showing mirror symmetry about the origin (Fig. 4.7). The sine waveform is just the opposite and is called anticentrosymmetric about the origin, because the wave on one side of the origin is the opposite of the other (Fig. 4.8). In a centrosymmetric waveform the phases of the Fourier components cannot take any arbitrary value. Each is a cosine function described by its amplitude and a sign

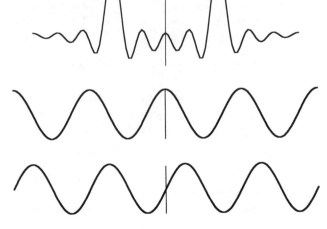

Fig. 4.6 A function centrosymmetric about the origin.

Fig. 4.7 A cosine wave is symmetric about the origin.

Fig. 4.8 A sine wave is antisymmetric, since the sine function is of opposite sign on either side of the origin.

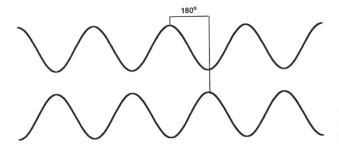

Fig. 4.9 A shift of phase of a wave by 180° is equivalent to reversing its sign.

Box 4.4 Cosine functions, symmetry, and phase angles

The cosine function is symmetrical about the origin: that is to say $\cos x = \cos(-x)$. The sine function is antisymmetric, since $\sin x = -\sin(-x)$ (see Figs 4.7 and 4.8). This means that the Fourier transform of any profile which is centrosymmetric about the origin (so that $\rho(x) = \rho(-x)$) can only contain cosine terms.

Remembering that the real part of $\exp(2\pi i h x + \phi) = \cos(2\pi h x + \phi)$, a symmetric wave profile can only be built up from waves of phase zero (an ordinary cosine wave) or π (a negative cosine wave, since $\cos(x + \pi) = -\cos x$). In this case, the structure factors $|F_h| \exp[i\phi_h]$ are real numbers, which will be $+|F_h|$ if the phase is zero or $-|F_h|$ if the phase is π.

(+ or −, according to whether there is a maximum or a minimum at the origin). The positive cosine function is a cosine wave with zero phase change, while the negative cosine function can also be seen as a wave, the phase of which is shifted by 180° (π radians) (Fig. 4.9). More detail is given in Box 4.4.

The orders h and $-h$: Friedel's law

An integer h is used to identify each of the different scattered waves forming the diffraction pattern, counting from 0 at the undiffracted beam which is parallel to the incident light. It is logical to label the orders on the other side of the unscattered beam with negative integers (Fig. 4.1). In normal diffraction, the intensities of diffraction of orders h and $-h$ are identical. F_h and F_{-h} have the same amplitude, but the signs of the phase angles of the scattered waves are opposite. This is known as Friedel's law (Box 4.5).

There is always a limit of resolution

There will always be a limit on the number of diffracted waves that can be observed. As h increases, the scattering angle ψ increases too (Fig. 4.1), but the distances travelled by rays scattered by adjacent units can never differ by more than the repeat distance a. This sets a maximum number of observable orders of diffraction. The maximum possible h is

Box 4.5 Diffracted orders h and $-h$: Friedel's law

Equation 1, Box 4.3 may be used to express the value of a structure factor with negative order $-h$:

$$F_{-h} = \int_0^1 \rho(x) \exp[-2\pi i h x] \, dx. \qquad (1)$$

By inserting $-h$ in the expression, the signs of all the exponents in the integral have been changed, but no other change has been made. This corresponds to the rule for forming a complex conjugate (Box 3.4). So F_{-h} may be written as:

$$F_{-h} = F_h^{\star} = |F_h| \exp[-i\alpha_h].$$

F_h and F_{-h} have the same amplitude, but the signs of the phases are opposite. This is known as Friedel's law.

Looking back at eqn 1, it is worth noting that this relationship between the phases of the Friedel pair, F_h and F_{-h}, depends upon the fact that the scattering density $\rho(x)$ is a real quantity. This remark may appear obscure now, but we shall return to it in Chapter 8.

less than a/λ. Thus if the repeat distance is 30 Å and the wavelength is 1.5 Å, only 20 orders of diffraction can be observed.

This means that, if we could use observations of the scattered waves to reconstruct the scattering profile, only this limited number of scattered waves can be included. The longer the wavelength λ, or the smaller the repeat distance a, the fewer orders of diffraction are accessible.

Summary so far

At this point it is worth reviewing the most important features presented so far.

1. With a monochromatic incident beam, a repetitive object scatters energy in specific directions, known as orders of diffraction.
2. The scattered waves have the same wavelength as the incident beam, but the scattered amplitude and phase depend on the distribution of scattering matter in the repeating unit.
3. The amplitude of the scattered wave may be measured by an X-ray detector, but X-ray detectors cannot observe the phase of an X-ray beam.
4. The amplitudes and phases of the scattered waves are related to the distribution of scattering matter by a well-known mathematical relationship, the Fourier transform. If the amplitudes and phases are known, Fourier transformation allows the distribution of scattering matter to be calculated.
5. The number of orders of diffraction which can be observed in a diffraction experiment is limited, and this limits the amount of detail that can be represented in the calculated distribution of scattering matter.

Building up an image from its Fourier components

If there are single equally spaced scattering points as in Figs 4.2 and 4.3, all the points scatter into the diffracted waves exactly in phase, with phase angle 0. The image of such a structure can be built up by adding the corresponding Fourier components, which are all cosine waves with peaks at the origin.

Figure 4.10 shows how the image develops as the second, third and fourth Fourier components are included, and the image including 10 Fourier terms. As more terms are included, the peaks representing the scattering points become sharper, and stand out more strongly from the background. If the scattering experiment was being conducted with X-rays, this image would show the electron density of the scattering array. Diffraction from metal crystals, which are often simple lattice arrays of metal atoms, are analogous to this in three dimensions.

Crystals of simple substances are often centrosymmetric. In centrosymmetric crystals the phase problem is replaced by a *sign problem* which is much simpler, because there are only two choices for the phase. If the crystal is centrosymmetric the same density of scattering matter exists at some point (x, y, z) and at the point $(-x, -y, -z)$ related to it by centrosymmetry. Figure 4.11 shows in one dimension the scattering from a repeated arrangement of points, which is irregular but symmetrical about $x = 0$. Because of this symmetry, all the Fourier components which build up its image are cosine waves. But because there is more than one scatterer in the lattice repeat, they have different amplitudes, and some of the waves are negative at the origin.

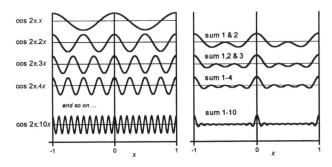

Fig. 4.10 Contributions to the scattering density from the first 10 orders of the diffraction from an equally spaced array of scattering points. These orders add to give an approximation to the scattering density.

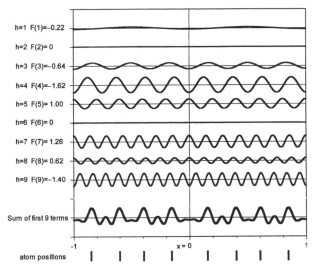

Fig. 4.11 Fourier summation based on a centrosymmetric structure with atoms at +0.15, 0.40, 0.60, 0.85. The individual Fourier terms are positive or negative cosine terms, all centrosymmetric about the origin, and the summation (here only the first nine terms) is also centrosymmetric.

Figure 4.12 shows the scattering from a non-centrosymmetric array of scattering points, regularly repeated to form a one-dimensional crystal. The waves representing the Fourier terms have their peaks in arbitrary positions, representing the corresponding phase angle. Although the majority of inorganic crystal structures are centrosymmetric, since the molecules which compose natural proteins and nucleic acids (and nearly all amino-acids and sugars) are never centrosymmetric, the determination of their structures inevitably involves phase angles other than 0° and 180°, as in this example.

Two kinds of wave have been mentioned and, to avoid confusion, it is important to distinguish these clearly. The X-rays scattered from the atoms in a particular direction form one kind of wave. This scattering is specified by the amplitude and phase of the scattered wave, relative to the incident wave. This quantity (specifying both amplitude and phase) is its structure factor.

A different kind of wave is used in building up the scattering density—in X-ray crystallography, the electron density $\rho(x)$. The distribution of scattering density is built up from a series of terms, forming a Fourier series. The distribution which produces the first order of diffraction is a wave of scattering density which repeats once within the repeating unit. A higher order of diffraction, h, corresponds to fluctuations in the scattering density which repeat h times within the repeating unit. Again, each of these has an amplitude, and a phase (which indicates where its peak is in relation to the origin).

Because the diffraction process creates a Fourier transform of the scattering structure, these two kinds of waves are related to each other. The X-rays scattered in a particular direction from the crystal have a particular amplitude and phase. The wave of electron density which contributes to the image of the structure has the same amplitude and phase. The rest of this chapter will make this relationship, between the scattered waves and the distribution of scattering matter, more familiar. This relationship is defined mathematically by the Fourier transform. But for practical purposes, you may consider this to mean the relationship between the distribution of scattering matter and the observed diffraction.

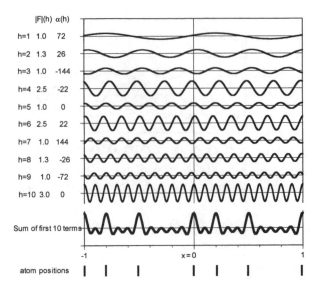

Fig. 4.12 Fourier summation based on a non-centrosymmetric structure with atoms at $x=0$, 0.2, 0.5. The Fourier terms have various phases, $\alpha(h)$.

Resolution

For an ideal simple diffraction grating (all the scattering material at regularly spaced points), every scattered wave is exactly in phase with the incident wave, all the scattered amplitudes are the same, and all the phases associated with the scattering factors are zero (Box 4.6). As the number of terms is increased, the image becomes more like the original object.

Because each scattered wave has the same amplitude and phase, all the Fourier terms generating the image are cosine waves of the same amplitude, with a maximum at the origin (Fig. 4.10). To generate the corresponding scattering density at some point x, we add up all the waves at this value of x. Away from the origin, the waves have a variety of phases, and they tend to cancel each other out. But at the origin, each wave is at its maximum value and the total is large. Figure 4.10 shows how the shape develops as further terms are included.

This demonstrates the effect of *resolution*. The Fourier image takes a more precise shape, as more orders of diffraction are included. But the precision is limited at each stage. Close to the origin every wave makes a positive contribution. It is positive over the half wavelength of the cosine, represented by $\cos(-90°)$ to $\cos(90°)$. The Fourier contribution to the image from order h has h peaks within the lattice repetition. If a total of n orders of diffraction are included, the nth Fourier contribution to the image has a wavelength $1/n$. The width of an image of a sharp line, when represented by a Fourier series of n orders, cannot be less than $1/2n$ of the repeat distance.

If there is a different distribution of scattering density, the various orders of diffraction have different amplitudes and phases. Figure 4.11 shows the scattering from a distribution of point scattering units (or 'atoms') which forms a centrosymmetric array. The orders of diffraction have various amplitudes. Each corresponds to a wave-like distribution of scattering density which is also centrosymmetric. These distributions are cosine waves of increasingly shorter wavelength, some of which have zero phase (maximum at the origin) and some have a phase of 180° (negative at the origin).

Figure 4.12 shows how the Fourier terms from a non-centrosymmetric arrangement of atoms builds up an image of the scattering distribution. Because the structure is not symmetric about the origin, the waves have various different phase angles.

Resolution is always limited, and any calculated scattering density $\rho(x)$ is always an approximation to the true density. As in Figs 4.10–4.12, the peaks representing the positions of the scattering atoms are at least $1/2n$ wide.

Box 4.6 Amplitudes and phases of scattering by a simple diffraction grating

A simple diffraction grating has scattering power S at regularly spaced points, but none between them. It may be represented by a scattering function $\rho(x)$ which is zero except where $x = 0$ or any integer. In the integral $\int_0^1 \rho(x) \exp[2\pi ihx] \, dx$, the only contribution arises when x is an integer, where $\exp[2\pi ihx] = \exp[0] = 1$, a real number.

Using Box 4.3, eqn 1, $F_h = \int_0^1 \rho(x) \exp[2\pi ihx] \, dx = S$, the scattering power of the material at the origin. This value is the same for all values of h. All the amplitudes $|F_h| = S$ and all the phases ϕ_h are zero.

Three useful Fourier transforms

So far, only scattering objects which are exactly repetitive have been discussed. In fact, the Fraunhofer diffraction from any scattering object can be analysed in just the same way, and it is found that the rays scattered in different directions represent the Fourier transform of the scattering object. The difference is that if the object is not repetitive, the scattering is not restricted to particular directions. Instead of orders of diffraction which can be numbered by integers h, the scattering function is a continuous one.

In order to see how the shape of a scattering object affects the diffraction pattern, it is useful to know the scattering properties of one appropriately shaped object.

We shall look at the Fourier transforms of three types of object. Box 4.7 indicates how to calculate them. The first of these represents the scattering from a uniform distribution

---

Box 4.7 Diffraction by non-repetitive objects

Diffraction may also occur from scattering matter which is not repetitive. In this case the scattering is not limited to individual orders of diffraction h. The distance from the centre of the diffraction pattern is labelled by a continuous variable ξ rather than by an integer h. The formula corresponding to eqn 1, Box 4.3 for Fraunhofer diffraction by a non-repetitive scattering array $\rho(x)$, in a direction ξ given by $\cos \psi = \xi\lambda$ is:

$$F(\xi) = \int_{-\infty}^{\infty} \rho(x) \exp[2\pi i\xi x] \, dx. \tag{1}$$

Note especially that the integral is now taken over an infinite range, and that the scattering factor may be non-zero for all values of ξ. Using this formula, the Fourier transforms of three simple one-dimensional functions may be calculated.

The Fourier transform of a 'top-hat function' (a wide block of scattering matter such that $\rho(x) = 1$ within a distance a of the origin) (Fig. 4.13) is obtained from eqn 1 as:

$$F(\xi) = \int_{-a}^{a} \exp[2\pi i\xi x] \, dx.$$

It is shown in books on integration that this reduces to

$$F(\xi) = F_0 \, 2\sin(2\pi\xi a)/2\pi\xi a \tag{1}$$

where F_0 is the scattering in the 'straight-through' direction.

The Fourier transform of two point scattering centres of density ρ_0, separated by a distance $2a$ (Fig. 4.14) may be obtained from eqn 1. It is:

$$F(\xi) = \rho_0 \, (\exp[2\pi i\xi a] + \exp[-2\pi i\xi a])$$

since ρ is zero except where $x = a$ or $-a$. This leads to:

$$F(\xi) = F_0 \cos(2\pi\xi a). \tag{2}$$

The Fourier transform of a Gaussian function of the form $\rho(x) = \exp[-a^2x^2]$ (Fig. 4.15) may be shown to be another Gaussian function:

$$F(x) = F_0 \exp[-2\pi\xi^2/a^2]. \tag{3}$$

All of these three Fourier transforms have a maximum value at the origin $\xi = 0$, and the transforms of the top-hat and Gaussian functions fade away to nothing at large ξ. The transform of two infinitesimal point-scattering centres is an infinite wave, but in practice all scattering centres have a finite size.

of scattering matter with a certain width. In one dimension this scattering function is called a 'top-hat' function because of its shape (Fig. 4.13). Its diffraction pattern has a peak at low scattering angles, which dies away at higher angles in an oscillatory fashion. This is the scattering from a single uniform scattering object of a certain size. In Fig. 4.13, the diagram on the left shows the variation of scattering density, and that on the right shows how the scattering varies in different directions. As the size of the object decreases, its scattering function becomes broader.

The second may be called a 'double-spike' function. It represents two scattering points. Each scatters in all directions, but the two scattered waves interfere with one another. In some directions they are exactly in phase, in others they cancel each other out. The result is a cosine wave of scattering amplitude (Fig. 4.14). The further apart the spikes are, the closer together the peaks of the cosine wave.

The third is a smoothly varying distribution (called a Gaussian function) of scattering matter. The density of scattering matter is high at the centre of the distribution, but dies away to nothing far from the centre. This distribution may be used to represent the density created by an atom that is vibrating, or whose position is blurred by thermal movements. An important property of a Gaussian function is that its Fourier transform is also a Gaussian (Fig. 4.15). A fuzzy object produces scattering that dies away at higher scattering angles. The fuzzier it is, the more rapidly the scattering dies away at higher angles.

The Fourier transforms are stated in algebraic form in Box 4.7. In all these cases, the narrower the diffracting objects are, the broader the width of the diffraction pattern. There is a reciprocal relationship between the dimensions of the diffracted object and those of the diffraction pattern.

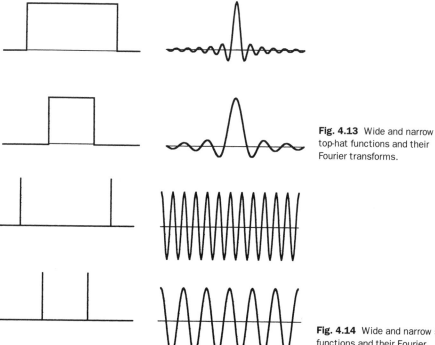

Fig. 4.13 Wide and narrow top-hat functions and their Fourier transforms.

Fig. 4.14 Wide and narrow spike functions and their Fourier transforms.

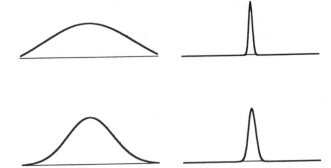

Fig. 4.15 Wide and narrow Gaussian functions and their Fourier transforms.

Two-dimensional diffraction

Two-dimensional optical diffraction patterns give the opportunity of presenting many properties of Fourier transforms in a graphic way. Henry Lipson, Charles Taylor, and Tom Wellberry refined techniques for making high-quality optical diffraction patterns originally designed by Lawrence Bragg. Beautiful diffraction patterns are published in their books, and some are reproduced here.

First, look at the two-dimensional diffraction pattern produced by a circular disc of scattering matter (Fig. 4.16a and g). The diffraction pattern is like a series of ripples, spreading out from the centre, but dying away. If a larger disc is used, the ripples come closer together (Fig. 4.16b and h). It is the two-dimensional equivalent of the top-hat function and its Fourier transform given in Fig. 4.13.

A pair of circular discs give similar diffraction patterns, except that they are overlapped by a pattern of waves (Fig. 4.16c, d, i, and j). This pattern of waves, spreading across the diffraction pattern, is a two-dimensional equivalent of the double-spike diffraction pattern shown in Fig. 4.14. In the same way as before, if the two scattering discs are moved further apart, the wave pattern becomes narrower (Fig. 4.16e, f, k, and l). The whole image superimposes the double-spike diffraction pattern and the top-hat diffraction pattern.

Instead of just two scattering points, there could be an array of scattering points arranged along a line. The line of points may run along the x-axis, spaced at intervals a. The diffraction produced by such an object has already been discussed in one dimension. It produces a series of equal orders of diffraction (Fig. 4.1). In two dimensions, the Fourier transform (or diffraction pattern) consists of a series of parallel lines, perpendicular to the x-axis (Fig. 4.17). The spacing of these lines along the x-axis proportional to $1/a$, again showing a reciprocal relationship to the dimensions of the scattering object. Each of the lines is associated with a particular order of diffraction h, which identifies the line.

A differently spaced line of points running in some other direction would create another series of parallel lines, perpendicular to the direction of spacing of the points (Fig. 4.18). These points may lie along a different axis y, spaced at a distance b. (Notice that the axial directions x and y are chosen to lie along the lines of points, and the lines are not necessarily perpendicular.) The spacing of the diffracted lines perpendicular to y will be $1/b$. The larger the spacing b, the closer together are the lines. Each of the lines represents a particular order of diffraction represented by a number which will be called k.

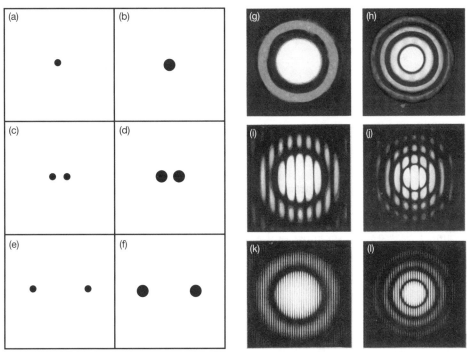

Fig. 4.16 Diffraction patterns from circular discs. (a) A small circular disc. (b) A larger circular disc. (c) and (d) Pairs of small and larger discs. (e) and (f) Pairs of small and larger discs, more widely spaced. (g)–(l) Corresponding diffraction patterns. (Reproduced by permission from Harburn *et al.* 1975.)

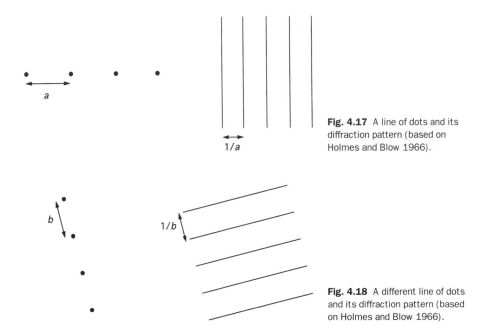

Fig. 4.17 A line of dots and its diffraction pattern (based on Holmes and Blow 1966).

Fig. 4.18 A different line of dots and its diffraction pattern (based on Holmes and Blow 1966).

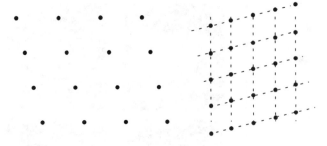

Fig. 4.19 A two-dimensional lattice and its diffraction pattern (based on Holmes and Blow 1966).

Figure 4.19 shows the diffraction from a regular two-dimensional arrangement of points forming a two-dimensional lattice. Thinking about it carefully, you can see that it is like the overlapping of the two diffraction patterns produced by the component lines of points. It is zero wherever either of the two separate diffraction patterns is zero, and has non-zero values only at points in the diffraction pattern where both the separate diffraction patterns are non-zero.

Convolution

Consider in detail a procedure by which a lattice may be built up from the two separate diffracting lines of points. Call the lines A and B, along the directions x and y, respectively, as shown in Figs 4.17 and 4.18. Line A has a series of equally placed points. If a line B is placed at every point where line A has a point, the complete lattice is generated (Fig. 4.19).

The operation carried out on lines A and B is called *convolution*. It can be specified generally as follows: consider two arrangements of diffracting density, A and B. At any point where A is non-zero, take this point as origin and generate the whole arrangement B, with a weighting according to the value of A at this point. Repeat this for every other point where A is non-zero. Add together all the densities generated by all these operations.

A more formal definition of convolution is given in Box 4.8.

Looking back at the two-dimensional diffraction results in Fig. 4.16, a convolution can be recognized. The scattering objects, c and e, composed of two discs are each the convolution of the single disk, a, with a double spike. (It doesn't matter which density distribution is called A and which is called B—their convolution has the same appearance.) The double discs, d and f, are related in just the same way to the single disc, b.

Now look at the diffraction patterns. The diffraction patterns in i, j, k, and l can be recognized as the product of the circular ripples in g and h with the wave diffraction patterns of the double spikes. This is an example of a general rule:

> If two diffracting densities are convoluted together, the resulting diffraction pattern is the product of the two individual diffraction patterns.

Because of the similarity of the Fourier transform operation and the inverse Fourier transformation, the rule can also be stated in an inverse sense:

> If two diffracting densities are multiplied together, the resulting diffraction pattern is the convolution of the two individual diffraction patterns.

Box 4.8 Definition of a convolution

Consider two arrangements of diffracting density $\rho_A(x)$ and $\rho_B(x)$. At any point x_A in $\rho_A(x)$, create a representation of $\rho_B(x)$ which takes x_A as origin, that is, $\rho_B(x-x_A)$. This is to be weighted by the value of $\rho_A(x_A)$, which may be written $\rho_A(x_A)$ $\rho_B(x-x_A)$. Adding up all the densities created by choosing every possible value of x_A is equivalent to integrating $[\rho_A(x_A)\,\rho_B(x-x_A)]$ over the possible values of x_A. Thus the convolution of $\rho_A(x)$ with $\rho_B(x)$ may be written:

$$\rho_C(x) = \int \rho_A(x_A)\,\rho_B(x-x_A)\,dx_A .$$

Two points are worth clarifying. The integral as stated is over an infinite range, but only needs to be taken over such a range of x_A where both $\rho_A(x_A)$ and $\rho_B(x - x_A)$ are non-zero, since the product is obviously zero elsewhere. The variable x_A, which identifies the point in A being used as the origin of ρ_B, is allowed to take all possible values in the integration, and is called a *dummy variable* by mathematicians. It is used in carrying out the mathematical operation, but disappears from the result.

Box 4.9 Fourier coefficients of a product of two repeating functions

Let the Fourier coefficients of two repeating functions $\rho_A(x)$ and $\rho_B(x)$ be $F_A(h)$, and $F_B(h)$. We shall now calculate the Fourier transform F_C of their product $\rho_A \cdot \rho_B$:

$$F_C(h) = \int \rho_A(x)\,\rho_B(x)\,\exp[2\pi ihx]\,dx.$$

First, $\rho_A(x)$ and $\rho_B(x)$ are each written out as a Fourier series:

$$F_C(h) = \int \left\{\sum F_A(h_A)\exp[-2\pi ih_A x]\right\}\left\{\sum F_B(h_B)\exp[-2\pi ih_B x]\right\}\exp[2\pi ihx]\,dx.$$

Since the functions are repetitive, the integral may be taken over one repeating unit $x=0$ to 1. Using the orthogonality rule (Box 3.7, eqn 1), the contribution to the integral of each A term in the first summation is only non-zero for the B term in the second summation for which $h-h_A-h_B=0$, or $h_B=h-h_A$. Taking these terms only,

$$F_C(h) = \int \left\{\sum F_A(h_A)\exp[2\pi ih_A x]F_B(h-h_A)\exp[2\pi i(h-h_A)x]\right\}\exp[2\pi ihx]\,dx$$

$$= \sum \left\{F_A(h_A)\,F_B(h-h_A)\right\}\int \exp[0]\,dx.$$

The F_A and F_B terms do not depend on x and have been taken outside the integration, and the total exponent is zero, so the integral is just 1. The Fourier transform of the product is:

$$F_C(h) = \sum F_A(h_A)\,F_B(h_A-h), \tag{1}$$

representing a convolution of the structure factors of A with those of B.

This box shows that 'the structure factors of the product of two densities are the convolution of the two corresponding sets of structure factors'.

Box 4.10 The density formed by the product of two sets of structure factors

In just the same way as Box 4.9, the inverse result can be calculated. The density formed by the product of two sets of structure factors may be written as $\rho_C(x)$, where

$$\rho_C(x) = \sum_h \{F_A(h)\, F_B(h)\, \exp[-2\pi ihx]\} \tag{1}$$

and the negative sign exists because this is an inverse Fourier transform (Box 3.9). Each structure factor can be replaced by a Fourier integral over the densities A and B:

$$\rho_C(x) = \sum_h \left\{ \int \rho_A(x_A)\, \exp[2\pi ihx_A] dx_A \times \int \rho_B(x_B)\, \exp[2\pi ihx_B] dx_B \right.$$
$$\left. \times \exp[-2\pi ihx] \right\}. \tag{2}$$

In eqn 2, x_A and x_B are dummy variables. Again, the only products in the integrals which will give a non-zero result are those for which $x_A + x_B - x = 0$, or $x_B = x - x_A$. Using the orthogonality rule, the product of integrals may be replaced by a single integral:

$$\rho_C(x) = \sum_h \left\{ \int \rho_A(x_A)\, \exp[2\pi ihx_A] \times \rho_B(x - x_A)\, \exp[2\pi ih(x - x_A)] dx_A \right\}$$
$$\times \exp[-2\pi ihx]. \tag{3}$$

For these terms the total exponent in the expression is zero, so

$$\rho_C(x) = \int \rho_A(x_A)\, \rho_B(x - x_A) \sum_h \exp[0]\, dx_A$$
$$= (2h_{max} + 1) \int \rho_A(x_A)\, \rho_B(x - x_A)\, dx_A. \tag{4}$$

Taking the product of the structure factors has led to the convolution of the two densities. The term $(2h_{max} + 1)$ is the total number of terms in the summation from $-h_{max}$ to h_{max}.

The final equation in this box and in Box 4.9 represent important general results about Fourier transforms. This box shows that 'the density corresponding to the product of two sets of structure factors is the convolution of the two densities represented by the sets of structure factors'.

Looking back at Figs 4.17–4.19, consider the dots in the left parts of the figures as diffracting objects. The array of dots in the right part of Fig. 4.19 may be recognized as the product of the density of the two sets of lines in Figs 4.17 and 4.18. The corresponding diffracting object at the left of Fig. 4.19 is the convolution of the left parts of Figs 4.17 and 4.18.

The corresponding mathematical relationships are derived in Boxes 4.9 and 4.10.

Simple lattices and their diffraction

We have seen in Fig. 4.19 how to develop a two-dimensional lattice by convolution of two lines of dots. Notice that if the line of diffracting dots is closer together, the parallel lines of their diffraction pattern are further apart. There is a reciprocal relationship between the spacings in the diffracting object and those of the diffraction pattern.

Convolution of the two lines of dots creates a lattice. The diffraction pattern of the lattice can be predicted from the convolution rule stated above. It will be the product of the two individual diffraction patterns. This product is easy to calculate, since at most points on the diffraction screen, one or other of the diffraction patterns is zero (i.e. blank). The only places where the product is not zero are the points where the lines cross. These points form a two-dimensional lattice.

This lattice diffraction pattern is shaped differently from the original lattice generated by convoluting two lines of dots. Its rows of dots are perpendicular to a line of dots in the diffracting object. The spacing of the rows of dots in the diffraction pattern depends reciprocally on the spacing of the lines of dots in the diffracting object.

These simple lattices are made up of *unit cells* which have dots at the corners, but none within the cell. The area of the unit cell in the diffracting object's lattice has a reciprocal relationship to the area of the unit cell of the diffraction pattern.

Crystal structures and their diffraction

Crystal structures are normally three-dimensional, but it is helpful now to consider what a two-dimensional crystal structure would be like. Some very simple crystals, like those of many metals, have atoms only at the lattice points. The diffraction patterns of these objects are similar to the diffraction pattern of a simple lattice of points shown in Fig. 4.19.

In crystal structures of complex molecules, scattering objects (electrons) are not close to the lattice points, but are spread out over the whole unit cell. How will this affect the diffraction pattern? It is easy to see, by carrying out another convolution operation.

Let us suppose we have a two-dimensional scattering object of characteristic shape (Fig. 4.20a). Its diffraction pattern is hard to calculate, but its intensity can easily be demonstrated in a diffraction experiment (Fig. 4.20b). Now suppose we had a crystal of these objects. What would its diffraction pattern be like?

'A crystal of these objects' means the convolution of a crystal lattice with one object (Fig. 4.21). This means that the diffraction pattern of this crystal will be the product of the diffraction pattern of the lattice with that of the individual object. Since the lattice diffraction pattern is zero except at the reciprocal lattice of points, the diffraction pattern of the crystal formed by convolution with this lattice is also zero except at the reciprocal lattice points. The intensity of the diffraction at one of these lattice points is proportional to the intensity of the duck's diffraction pattern at the corresponding point. This can be described as 'sampling' the duck's diffraction pattern at the reciprocal lattice points.

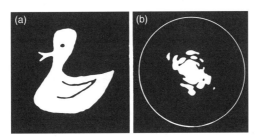

Fig. 4.20 (a) A duck-shaped scatterer; (b) its diffraction pattern. (Reproduced by permission from Taylor and Lipson 1964.)

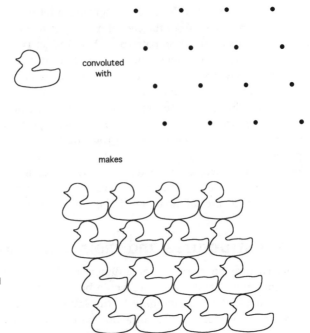

Fig. 4.21 A lattice of duck-shaped objects may be produced by convoluting the duck shape with a lattice (based on Holmes and Blow 1966).

Resolution in two dimensions

We already saw in one dimension that the number of orders of the diffraction pattern which can be observed (using radiation of a given wavelength) is limited. The same applies in two dimensions. Compared to an unlimited Fourier transform which could be computed in two dimensions, the optical diffraction pattern is observable only within an area of limited radius. What effect will this have on the image?

The observable diffraction pattern consists of the ideal two-dimensional transform, with the area outside the observable region screened off. This is the same as multiplying the duck's diffraction pattern by a disc function of given size. In Figs 4.22 and 4.23, two screened-off diffraction patterns are illustrated, together with the image that can be generated from them. The larger the amount of the diffraction pattern that can be observed (the wider the disc), the more detail is generated in the image.

These less-detailed images can be considered in another way. They are the convolution of the original scattering object shown in Fig. 4.20 with the diffraction pattern of the screening disc. If the screening disc is wider, more of the original diffraction pattern is included, and the image resembles the original more closely. Looking at it in another way, the larger the amount of diffraction that can be included, the wider the disc, the sharper the peak of its diffraction pattern, and the less the convolution smears the image.

Figures 4.22 and 4.23 show what happens when the observable diffraction pattern is restricted in size by physical constraints. In imaging a real molecule, other effects limit the resolution. The atoms of the molecule are in constant thermal vibration. Over a period of time, the average electron density of an atom may be represented reasonably well by a Gaussian function (Fig. 4.15) of appropriate radius. We may say that the true electron density of each atom is convoluted by a Gaussian function which represents thermal vibrations. So its diffraction pattern is multiplied by the Fourier transform of this Gaussian function,

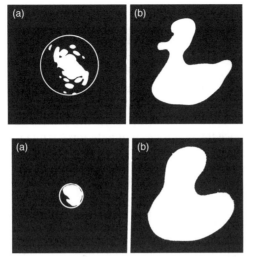

Fig. 4.22 Diffraction pattern shown in Fig. 4.20 screened off beyond a given radius (the 'resolution') (a), and the image that is generated from it (b). (Reproduced by permission from Taylor and Lipson 1964.)

Fig. 4.23 Diffraction pattern shown in Fig. 4.22 screened off at a smaller radius (a), and the image that is generated from it (b). (Reproduced by permission from Taylor and Lipson 1964.)

which, as shown in Fig. 4.15, is also a Gaussian function. The wider the cloud representing the vibrations, the narrower is its Fourier transform. Modification of the diffraction pattern by a Gaussian function reduces the intensity of the outer parts of the pattern, ultimately reducing it to the point where the intensity cannot be measured. The larger the amplitude of the molecular vibrations, the smaller is the region within which measurements are possible.

In practice, in a macromolecular crystal, a similar effect places an even more important constraint on the size of the measurable diffraction pattern. At any instant, the contents of different unit cells of the crystal are not identical. This is partly because of thermal atomic vibrations. But in most crystals, more serious disorder exists because of crystal imperfections. This disorder, also, can often be represented pretty well by a Gaussian disturbance of the contents of an 'average' unit cell; and leads to a Gaussian modification of the observable diffraction pattern, which is usually more severe than that caused by thermal vibrations. Practical details will be taken up in Chapter 5.

These limitations on observation of a diffraction pattern constrain the sharpness, or 'resolution' of the best image that can be obtained from it. To put a number on the resolution, we refer to the spacing of a row of dots, the first-order diffraction of which would lie on the limit of measurement of the diffraction pattern. If diffraction can not be measured beyond the point where a lattice row spaced at 2 Å would give its first-order diffraction, crystallographers say that the resolution is 2 Å. This means that if one unit cell dimension is 50 Å, and diffraction is measured for 25 orders in this direction, the resolution would be 2 Å.

The larger the domain of the diffraction pattern that can be observed, the finer the detail of the image. There is a direct relationship between the amount of diffraction data recorded, and the sharpness of the image. As the available diffraction measurements are extended, crystallographers say that they are increasing the resolution, or working at higher resolution. Of course, the distance (in, say, Ångstroms) representing the resolution limit becomes smaller.

The Patterson function

There is an important crystallographic technique which can be understood easily in terms of convolution. It was invented in 1934 by Lyndo Patterson, and is always linked with his name. It provided an important new way of working back from the diffraction pattern to the crystal structure, avoiding the phase problem. It was invaluable for the crystal structures that could be studied at that time.

As already explained, the Fourier transform of a repetitive scattering object represents a set of orders of diffraction, each of which is described by an amplitude and a phase. The amplitude of a scattered wave can be measured from the intensity of the diffracted X-ray beam, which is proportional to the square of the amplitude of the corresponding wave. The intensity can be observed directly. The phase of an X-ray beam is not directly observable.

Patterson therefore asked himself what could be learned about a scattering structure from observation of the intensities. Since the intensities represent the square of the amplitudes, they can be thought of as the product of the scattering with itself. The Fourier transform of the intensities, known as the Patterson function, would then be the convolution of the scattering structure with itself. Box 4.11 presents these statements with more precision.

Box 4.11 The Patterson function

The Patterson function, loosely called the self-convolution of a structure, is more precisely the convolution of a structure $\rho(x)$ with its inverse $\rho(-x)$. It is the set of vectors between the atoms of a structure, appropriately weighted. To emphasize that the Patterson function displays a set of vectors, it is usually written as a function of u, u being a symbol to represent a vector, in contrast to x to represent a coordinate.

The Patterson function may be calculated as:

$$P(u) = \sum_h |F_h|^2 \exp[-2\pi i h u].$$

The beginning of Chapter 3 explains that the energy carried by a wave, called the intensity, is proportional to the square of the amplitude. Therefore the Patterson function is the Fourier transform of the intensities, quantities that can be measured directly.

The scattered X-ray intensity $|F_h|^2$ may be written as $F_h . F_h^*$ following Box 3.4, eqn 1.

$$P(u) = \sum_h \{F(h)F^*(h) \exp[-2\pi i h u]\}. \tag{1}$$

The argument of Box 4.10 may be followed through, starting from this definition. Note that the complex conjugate $F^*(h) = \int \rho(x) \exp[-2\pi i h x] dx$. Using x_A and x_B as dummy variables, Box 4.10, eqn 2 becomes:

$$P(u) = \sum_h \left\{ \int \rho(x_A) \exp[2\pi i h x_A] dx_A \times \int \rho(x_B) \exp[-2\pi i h x_B] dx_B \right.$$

$$\left. \times \exp[-2\pi i h u] \right\}. \tag{2}$$

In this case, non-zero terms arise only if $x_B = x_A - u$, leading to

$$P(u) = \sum_h \{ \int \rho(x_A) \exp[2\pi i h x_A] \times \rho(x_A - u) \exp[-2\pi i h (x_A - u)] dx_A \}$$

$$\times \exp[-2\pi i h u]. \tag{3}$$

$$P(u) = \int \rho(x_A) \rho(x_A - u) \sum_h \exp[0] dx_A$$

$$= (2h_{max} + 1) \int \rho(x_A) \rho(-[u - x_A]) dx_A. \tag{4}$$

The negative sign before the brackets means the convolution is that of $\rho(x_A)$ with $\rho(-x_A)$. Thus the Patterson function is the convolution of the electron density with its inverse.

'Convolution of the scattering structure with itself' is not so hard to visualize if the structure is composed of point atoms. Take an idealized benzene molecule as an example, represented by six points forming a regular hexagon (Fig. 4.24). Convolution with itself means taking six such regular hexagons, and putting them so that a different one of the six points lies at the origin. Figure 4.25 shows the six hexagons, and the array of weighted points which is the self-convolution of this structure.

Each hexagon has one point at the origin, so there is a total weight of 6 units at the origin. Two of the hexagons generate a point one bond-length away from the origin, in each of six directions, giving a double weight to this point. There are six more points a little further from the origin, each lying on a corner of two different hexagons, so these points also have double weight. Altogether there are 12 points that have a weight of 2 units. Each of the hexagons generates another point unique to it, creating 6 points of unit density.

Another way of looking at it is to think of all the vectors between atoms in the benzene molecule. Every atom has a vector of zero length to itself (such as AA), so there are six such vectors at the origin. There are bonds between adjacent atoms (such as AB) in each of six directions, but in each case there are two parallel bonds (AB and ED), so these six vectors each have a weight of 2 units. There are also vectors to second-nearest neighbours in each of six directions (such as AC), but in each case there are two such vectors (AC and FD) so these six points also have a weight of 2 units. Finally, there are vectors diametrically across the molecule (such as AD) in each of six directions, giving six points of single weight.

Whichever of these two methods is used, there are altogether 36 (6^2) units of Patterson density, of which six are at the origin. In general, a molecule of N atoms generates a Patterson density with a total of N^2 units, of which N are at the origin and the rest are elsewhere.

The Patterson function generated from the scattering by crystals of benzene can be calculated directly from the observed intensities. It shows this characteristic distribution of Patterson density which is easily recognized (Fig. 4.26). This method led directly to molecular structure determination for many simple molecules.

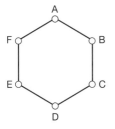

Fig. 4.24 A benzene molecule.

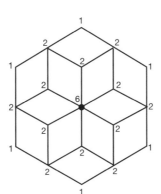

Fig. 4.25 Each atom of the benzene molecule has been placed in turn at the origin, and the hexagon representing the molecule has been drawn. The numbers state the number of hexagons that coincide at each vertex.

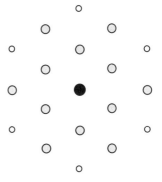

● 6   ○ 2   ○ 1

Fig. 4.26 The Patterson function of a benzene molecule.

The Patterson function summarizes all the data available from the intensities of the diffraction pattern. Many smaller crystal structures were 'solved' directly from inspection of the Patterson function.

Three dimensions

This chapter has shown in one and two dimensions that:

(1) a regular array of scattering units diffracts to form a regular array of scattered beams;
(2) if the observed diffraction is restricted to a certain distance from the origin (the top-hat or disc function), the image of the scattering units is broadened;
(3) if the scattering objects are smeared around their ideal position (a Gaussian function), the diffraction pattern becomes weaker at higher scattering angles;
(4) if the scattering objects have an internal structure, the scattering from an array of objects samples the Fourier transform of that internal structure.

Most of these points can be extended easily into a third dimension (Box 4.12). For example, a three-dimensional lattice may be generated by convoluting a two-dimensional lattice with a uniformly spaced row of points that do not lie in the plane. The three-dimensional lattice is composed of unit cells, the volume of which depends on the spacings and directions of the three lines of points that generate it. A three-dimensional crystal of molecules can be generated by convoluting the three-dimensional lattice with one copy of the molecule.

Box 4.12 Three dimensions

The notation presented so far can be extended easily into a third dimension. The extension of Fourier analysis to three dimensions requires a three-dimensional scattering function $\rho(x, y, z)$, and a three-dimensional array of Fourier terms identified by coefficients (h, k, l). A three-dimensional lattice may be generated by convoluting a two-dimensional lattice with another uniformly spaced row of points which do not lie in the plane.

In three dimensions, the equations of Box 4.3 would be written:

$$F_{hkl} = |F_{hkl}| \exp[i\phi_{hkl}]$$
$$= \int_0^1 \int_0^1 \int_0^1 \rho(x, y, z) \exp[2\pi i(hx + ky + lz)] \, dx \, dy \, dz, \tag{1}$$

and

$$\rho(x, y, z) = \sum_{hkl} F_{hkl} \exp[-2\pi i(hx + ky + lz)]. \tag{2}$$

It is important to remember that throughout the boxes of this chapter the symbol x has been used to represent a fraction of the repeat length, not to be confused with the Cartesian coordinate $X = xa$. The same applies to the three-dimensional formulae where $x, y,$ and z are fractions of the lattice translations $a, b,$ and c. As we have seen, in two- and three-dimensional lattices, the lattice translations are not necessarily perpendicular to each other, and neither are the fractional lattice coordinates $x, y,$ and z.

Similar relations exist between the Fourier transforms of the various three-dimensional objects, and the corresponding two-dimensional objects already discussed.

The important difference is that in three dimensions the Fourier transform is not strictly a diffraction pattern. Diffraction by three-dimensional arrays of objects has an additional complication which will be discussed in the next chapter.

Further reading

Especially recommended amongst presentations of diffraction in many physics textbooks:
Lipson, S.G., Lipson, H. and Tannhausen, D.S. (1995). *Optical physics* (3rd edn). Chapters 4 and 8. Cambridge University Press, Cambridge.

Diffraction principles are clearly presented in two dimensions by:
Taylor, C.A. and Lipson, H. (1964). *Optical transforms* (2nd edn). Bell, London.

Beautiful two-dimensional examples of diffraction can be seen in:
Harburn, G., Taylor, C.A. and Wellberry, T.R. (1975). *Atlas of optical transforms*. Bell, London.

5

Diffraction by crystals

In Chapter 4 many two-dimensional examples were shown, in which a diffraction pattern represents the Fourier transform of the scattering object. When a diffracting object is three-dimensional, a new effect arises.

In diffraction by a repetitive object, rays are scattered in many directions. Each unit of the lattice scatters, but a diffracted beam arises only if the scattered rays from each unit are all in phase. Otherwise the scattering from one unit is cancelled out by another.

In two dimensions, there is always a direction where the scattered rays are in phase for any order of diffraction (just as shown for a one-dimensional scatterer in Fig. 4.1). In three dimensions, it is only possible for all the points of a lattice to scatter in phase if the crystal is correctly oriented in the incident beam.

The amplitudes and phases of all the scattered beams from a three-dimensional crystal still provide the Fourier transform of the three-dimensional structure. But when a crystal is at a particular angular orientation to the X-ray beam, the scattering of a monochromatic beam provides only a tiny sample of the total Fourier transform of its structure (Fig. 5.1).

In the next section, we are going to find what is needed to allow a diffracted beam to be generated. We shall follow a treatment invented by Lawrence Bragg in 1913. Max von Laue, who discovered X-ray diffraction in 1912, used a different scheme of analysis; and Paul Ewald introduced a new way of looking at it in 1921. These three methods are referred to as the Laue equations, Bragg's law and the Ewald construction, and they give identical results. All three are described in many crystallographic text books. Bragg's method is straightforward, understandable, and suffices for present needs.

Fig. 5.1 'Still' X-ray photograph (courtesy of Emmanuel Saridakis).

Lawrence Bragg wrote about his discovery as follows:

I had heard J.J. Thomson lecture about ... X-rays as very short pulses of radiation. I worked out that such pulses ... should be reflected at any angle of incidence by the sheets of atoms in the crystal as if these sheets were mirrors. ... It remained to explain why certain of the atomic mirrors in the zinc blende [ZnS] crystal reflected more powerfully than others. Pope and Barlow had a theory that the atoms in simple cubic compounds like zinc blende were packed together, not like balls at the corners of a stack of cubes, but in what is called cubic close packing, where the balls are also at the centre of the cube faces. I tried whether this would explain the anomaly—and it did! ... These were the first crystals to be analysed by X-rays (... June 1913).   (Bragg 1943)

Bragg's law—first form

The problem is to find how the scattering from every lattice point can be in phase. The first part of Bragg's explanation is to show how rays scattered at the lattice points in one plane all travel the same distance to a scattered wavefront. If they travel the same distance, they are exactly in phase. This happens when this plane of lattice points behaves as a mirror. Suppose a parallel beam of X-rays falls on a plane of scattering points which lie in a plane, and which scatter X-rays in all directions. In the direction corresponding to reflection using this plane as a mirror, the X-rays travel exactly the same distance to form a wavefront, for all positions in this plane of the lattice (Fig. 5.2).

In this figure, the rays that compose the incoming wavefront start off in phase, and arrive at the reflected wavefront in phase, so they add up to a strongly reflected beam. In other directions, rays scattered from different parts of the mirror travel different distances, so the phase differences between the scattered waves cause them to cancel each other out. This is one way of explaining how a mirror works.

Once this idea has been grasped, the second part of Bragg's argument is quite easy to follow. Think of two parallel half-silvered mirrors. Some light is scattered by the first mirror, and some passes through it and is scattered in the second. The two mirrors only scatter in phase if the difference of path length, between scattering in the first plane and the second, is a whole number of wavelengths. In Fig. 5.3 the difference is one wavelength, half a wavelength more to get to the mirror, and another half wavelength after reflection. Figure 5.4 shows that the extra path difference for rays scattered at the second mirror is $2d \sin \theta$, where θ is the 'glancing angle', which is the angle between the incident beam and the mirror, and d is the spacing between the mirrors.

In a crystal, any plane of lattice points can act as a mirror. The crystal lattice creates many parallel lattice planes. The condition for two planes to scatter in phase depends on three quantities:

- the wavelength λ of the X-rays;
- the spacing d between the planes;
- the glancing angle θ.

As in Fig. 5.4, the two planes will scatter in phase if this path difference $2d \sin \theta$ is a whole number of wavelengths, say n:

$$n\lambda = 2d \sin \theta. \tag{5.1}$$

If these two adjacent planes scatter in phase, the lattice symmetry means that this will apply to the second plane and a third parallel plane, and so on, to every lattice point of the crystal. Eqn 5.1 is the first way of expressing Bragg's law. We shall soon come to a second, more informative, form of the law.

In Fig. 5.3, the extra path was one wavelength, corresponding to $n = 1$ in eqn 5.1. If the extra path between adjacent lattice planes were two wavelengths, as illustrated in Fig. 5.5, the reflected beams from each lattice plane are still in phase. This example corresponds to $n = 2$ in eqn 5.1.

Bragg's idea that every order of diffraction can be considered as reflection in a set of lattice planes pervades X-ray crystallography. A diffracted beam is often referred to as a 'Bragg reflection' or just as a 'reflection'.

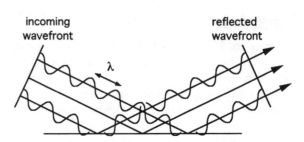

Fig. 5.2 For reflection in this plane as a mirror, the rays have to travel exactly the same distance to form a wavefront, no matter where they are scattered.

Fig. 5.3 Rays scattered in a pair of parallel mirrors so that the reflected rays are in phase. The incident and the reflected rays each have an extra distance $\lambda/2$ to travel.

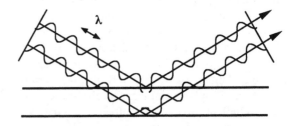

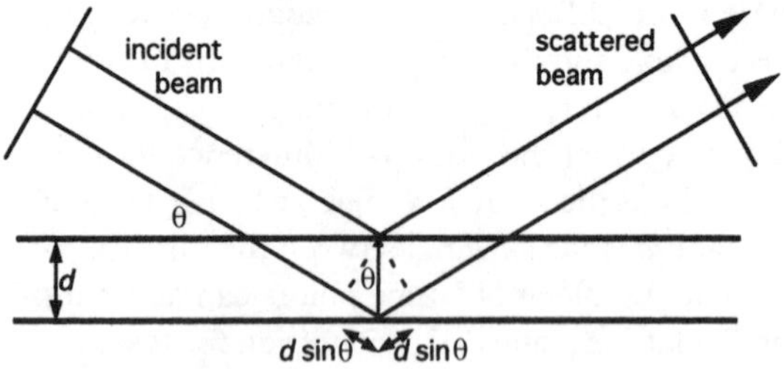

Fig. 5.4 The extra path for X-rays scattered at the second plane is $2d \sin \theta$.

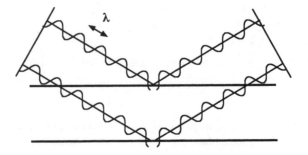

Fig. 5.5 In this example reflection at the second mirror introduces an extra path length of 2 wavelengths, as can be confirmed by counting the wavelengths in each of the beams.

Lattice planes

There are many possible lattice planes associated with a lattice. If you ride past trees in a regularly spaced plantation, you will pass through many directions where rows of trees are in line. In just the same way, there are many possible directions for a plane through a lattice, such that lattice points lie on the plane.

Figures 5.6–5.10 present several directions of lines which pass through the points of a two-dimensional lattice. Each shows a pair of parallel lines, both passing through lattice points. The second line has been chosen to be as close as possible to the first, while still passing through lattice points. If one repeated drawing parallel lines at this spacing, all the points in the lattice could be covered.

One unit cell is outlined in each diagram. One of the lattice lines passes through the origin of the unit cell, at its bottom left-hand corner. In the first example, Fig. 5.6, the lines are parallel to the b axis of the lattice. One line is displaced from the next by a distance equal to the a axis. In Fig. 5.7, the second line passes through opposite corners of the unit cell. In Fig. 5.8, the second line passes through one corner of the cell, and through the half-way point of the opposite side. If the lines pass through lattice points, these intersections must always occur at an exact fraction of the cell edge.

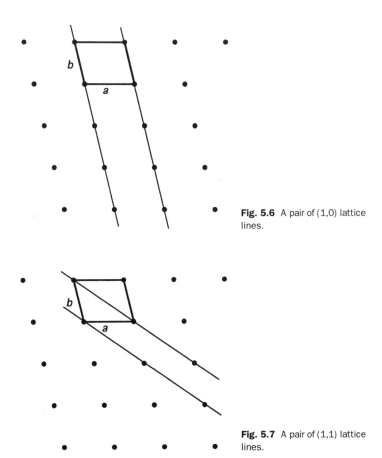

Fig. 5.6 A pair of (1,0) lattice lines.

Fig. 5.7 A pair of (1,1) lattice lines.

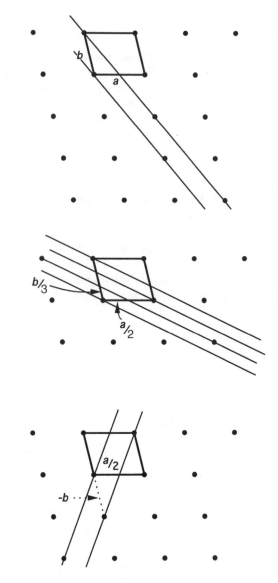

Fig. 5.8 A pair of (2,1) lattice lines.

Fig. 5.9 A pair of (2,3) lattice lines.

Fig. 5.10 A pair of (2,−1) lattice lines.

In Fig. 5.9 another set of such lattice lines is shown. In this case, the second lattice line cuts the cell edge at exactly half of the a spacing and one-third of the b spacing. Then, if the next parallel line is displaced by one-third of the cell along b, it will pass through a further set of lattice points. A series of equally spaced parallel lines will cover all the points in the lattice.

The fractional distance along the unit cell edges between pairs of lattice lines can be used to label a particular set of lattice lines. The label consists of two integers h and k, representing the fraction along a and the fraction along b. In Fig. 5.9, h is 2 and k is 3. The lattice lines are called the (2,3) lattice lines.

In Fig. 5.5, the lattice lines run parallel to one of the axes. However far you go, the second line will never cut the b axis of the cell. The distance can be considered infinite and the k index is assigned as zero. This diagram shows (1,0) lattice lines.

The positive directions along the axes are drawn with a going to the right and b going upwards. In Fig. 5.10 an example is shown where one of the axes is cut by the lattice line in a negative direction. This set of lattice lines is called the $(2,-1)$ lattice lines. When the unit cell is oblique, as illustrated, these lattice lines are at a different spacing from the (2,1) lines.

Figure 5.11 brings several of the examples together in one diagram. It shows just two lattice lines for each case, one of them passing through the origin at the lower left of the cell, and the next line adjacent to it. It illustrates how, as the indices become larger, the lattice lines are spaced more closely together.

Figure 5.12 shows another possible choice for h, k, in this case the (4,6) lattice lines. They run in the same direction as the (2,3) planes (shown as full lines), but their spacing is halved. The dashed lines (the (4,6) lines which are not also (2,3) lines) do not pass through any lattice points. Even so, the full set of (4,6) lines covers every point in the lattice. You will see presently why lattice lines like this are useful.

Exactly the same type of construction can be made in three dimensions (Fig. 5.13). If one plane passes through a point of the three-dimensional lattice, an adjacent plane must

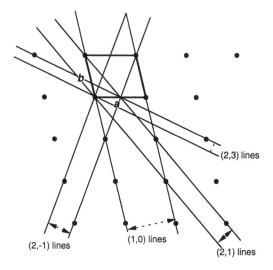

Fig. 5.11 As the lattice indices become larger, the lattice spacings get shorter.

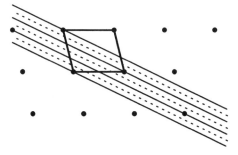

Fig. 5.12 The (4,6) lattice lines. The dashed lines are (4,6) lines, but pass through no lattice points.

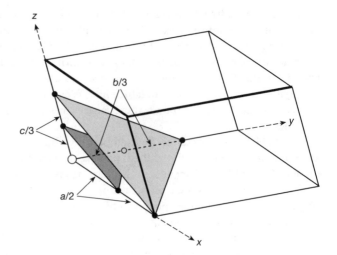

Fig. 5.13 The same kind of construction in three dimensions. This example shows the (2,3,3) planes.

cut through the three unit cell edges at exact fractions, say $1/h$, $1/k$, $1/l$ of the three cell edges, where h, k, l are integers. All the possible sets of planes which cover all the lattice points can be described by different values of h, k, and l (values may be negative or zero as already described). The planes in Fig. 5.13 are the (2,3,3) planes. (This notation is often abbreviated to (233) if all the indices are less than 10.)

The set of three integers, h, k, l, therefore specifies a particular set of lattice planes. A different set of integers specifies a different set of planes. Any of these sets of planes may be the ones that act as mirrors in the diffraction effect. Eqn 5.1 can be modified to specify which particular set of planes is in the reflecting position:

$$n\lambda = 2d_{hkl}\sin\theta. \tag{5.2}$$

Bragg's law—second form

Let us now go back to the (4,6) lines of Fig. 5.12. The same kind of lattice planes can be imagined in a three-dimensional lattice, say the (466) planes of Fig. 5.14. Compared to the (233) planes, their spacing, d_{466}, is halved. Although some of the planes go through no lattice points, all the lattice points lie on one of the equally spaced parallel planes.

Figure 5.5 illustrated two mirrors, spaced so that there were two wavelengths path difference, for waves reflected in the second mirror. This arrangement allows all the reflected rays to be in phase, just as well as that shown in Fig. 5.3.

Suppose n in eqn 5.2 were 2, so that there are 2 wavelengths of path difference between reflections in successive planes at spacing d_{hkl}. If we chose the set of planes $(2h,2k,2l)$, their d spacing is exactly halved and would give exactly one wavelength of path difference. This is illustrated in Fig. 5.15. In eqn 5.2, as written above, if $n = 2$ then

$$2\lambda = 2d_{hkl}\sin\theta,$$

and therefore

$$\lambda = 2d_{2h,2k,2l}\sin\theta.$$

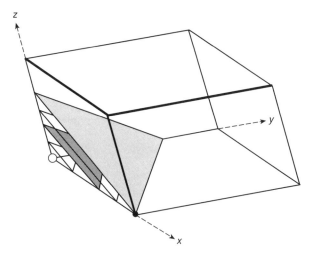

Fig. 5.14 The (4,6,6) planes are at exactly half the spacing of the (2,3,3) planes.

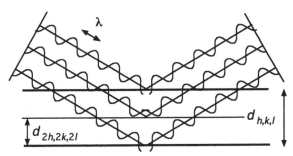

Fig 5.15 The set of planes (2h,2k,2l) have exactly half the spacing of the (h,k,l) planes. If (h,k,l) gives two wavelengths of path difference for reflection in adjacent planes, (2h,2k,2l) gives one wavelength of path difference.

Work through the same argument for $n = 3$. This will show that for any value of n, we can always modify our choice of h, k, l, so as to write the equation in the form:

$$\lambda = 2d_{hkl}\sin\theta. \tag{5.3}$$

All the possible types of diffraction from the three-dimensional lattice are covered by all the possible choices of h, k, l. This form of Bragg's law in eqn 5.3 is the standard way of expressing it.

Discussion of Bragg's law

By this slight generalization, h, k, l can be used to enumerate all the possible types of diffraction from a crystal. The numbers h, k, l are called the *indices* of a particular order of diffraction. A spacing d_{hkl} is associated with each set of indices.

In the most usual diffraction method, the X-ray beam is monochromatic, that is to say, λ is fixed. Then for any particular value of d_{hkl} a value of θ is specified by Bragg's law. As seen in Fig. 5.3, θ is the glancing angle at which the X-rays fall on the plane h, k, l. Eqn 5.3 can be rearranged to

$$\sin\theta = \frac{\lambda}{2d_{hkl}}. \tag{5.4}$$

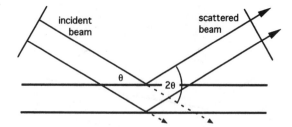

Fig. 5.16 The total deviation of the scattered beam from the incident X-ray beam is twice the glancing angle θ.

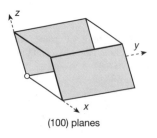

(100) planes

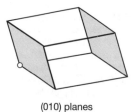

(010) planes

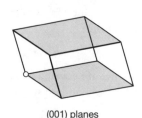

(001) planes

Fig. 5.17 (100) planes, (010) planes, and (001) planes.

This form emphasizes that the angle θ depends upon d_{hkl}. The total deviation of the scattered beam from the incident X-ray beam is 2θ (Fig. 5.16). The larger the deviation angle, the larger is sin θ, and the shorter the spacing between the corresponding set of lattice planes. A small spacing corresponds to finer detail, or improved resolution. Thus the part of the diffraction pattern near the undeviated beam (θ small) is related to coarser details of the scattering structure, and the further we go from the direct beam (θ increasing), the finer the detail or the 'higher' the resolution.

This is entirely consistent with what has been shown already in one and two dimensions. Since sin θ cannot exceed 1, the smallest d_{hkl} corresponding to the finest resolution possible is λ/2. At this value of θ, the incident beam is at 90° to the reflecting plane of atoms, so the diffracted beam is sent straight back along the incident direction. (This would make it difficult to observe, so in practice θ has to be a bit less than 90°!)

Determination of unit cell dimensions

The *d*-spacing associated with any reflection can be determined directly by observation of its θ angle and use of Bragg's law. The simplest cases are the θ angles associated with the reflections for which (*hkl*) are (100), (010), and (001). These are the planes that pass through two opposite faces of the unit cell (Fig. 5.17), and give the perpendicular spacing between them. (It is important to notice that unless the crystal axes are perpendicular, these spacings are *not* the unit cell dimensions *a*, *b*, *c*, and some geometry is still required to determine them.)

The angles between the crystal axes, where these are not fixed by the symmetry, are determined by measuring the angular relationships between reflecting positions of the crystal for different reflections, and by the directions of the reflected beams.

Calculation of three-dimensional Fourier transforms

In crystallography, the coordinates x, y, z present the three-dimensional coordinates of a point in the unit cell. These coordinates represent fractions of the unit cell edge (not Ångstroms or any other fixed length unit). Box 5.1 presents, in three-dimensional form, the equations for calculating the structure factors from the electron density, and the inverse transformation, corresponding to those in Box 4.3. Box 5.2 presents an alternative formula which can be used to calculate the structure factors from the positions and properties of all the atoms in the unit cell.

In Chapter 4 we considered the smearing of the density of an atom by its thermal vibration or by crystal disorder, and mentioned that this could be represented well by a Gaussian function. As a result, the diffraction pattern of the atom is modified by its Fourier transform, also a Gaussian function, which weakens the diffraction at larger θ angles. The magnitude of this effect depends upon the B factor defined in Box 5.2.

Box 5.1 Three-dimensional formulae for structure factors and for electron density

Eqn 1, Box 4.3 gives the structure factor formula in one dimension:

$$F_h = \int_0^1 \rho(x) \exp[2\pi i h x] \, dx.$$

The formula in three dimensions is similar:

$$F_{hkl} = |F_{hkl}| \exp[i\phi_{hkl}]$$

$$= V \int_0^1\int_0^1\int_0^1 \rho(x,y,z) \exp[2\pi i(hx + ky + lz)] \, dx\,dy\,dz. \tag{1}$$

Taking the differences between these equations in turn, the factor V represents the volume of the unit cell. It arises because ρ is a density expressed as electrons per unit volume, while the structure factor F is normally expressed as electrons per unit cell. Three integrations are taken over 0 to 1 in each of x, y, and z, so as to extend it over the whole unit cell. The phase factor corresponding to $2\pi h x$ is $2\pi(hx + ky + lz)$ in three dimensions.

The inverse Fourier transform, eqn 2, Box 4.3 becomes

$$\rho(x,y,z) = \frac{1}{V}\sum_{hkl} F_{hkl}\exp[-2\pi i(hx+ky+lz)]. \tag{2}$$

The summation symbol indicates summation over all combinations of h, k, and l. It is sometimes written using three separate summation signs.

To keep the notation compact, it can be further simplified by writing x for (x,y,z), h for hkl and $h \cdot x$ for $hx + ky + lz$. The quantities x and h are formally vectors, but for our purposes they are just an abbreviation. Eqn 2 then becomes:

$$\rho(x) = \frac{1}{V}\sum_{h} F(h) \exp[-2\pi i h \cdot x]. \tag{3}$$

Using similar notation, eqn 1 becomes:

$$F_{hkl} = V \int \rho(x) \exp[2\pi i(h \cdot x)] \, dV. \tag{4}$$

Box 5.2 The structure factor as a sum of scattering by all the atoms

The electron-density function $\rho(x)$ represents the electron density (the scattering power) of all the atoms in the crystal unit cell. Instead of an integral over the density, eqns 1 and 4 of Box 5.1 can be written using a sum of the scattering due to each of the N atoms in the cell. This must take account of the different shape of the electron density of each atom, and for the disorder of the structure due to thermal vibrations and crystal disorder. We write:

$$F_{hkl} = \sum_{N} f_i \exp[2\pi i(hx_i + ky_i + lz_i)] \exp[-B_i \sin^2\theta/\lambda^2]. \tag{1}$$

The quantity f_i represents the scattering by the ith atom and is called the *atomic scattering factor*. It varies with d_{hkl}, to form the Fourier transform of the electron-density distribution within the particular atom. At low resolution it equals the number of electrons in the atom, but reduces at higher resolution according to the shape of the electron distribution around the atom. It is usual to consider that f_i is the same for every atom of the same type, ignoring the distortion of electron density into bonding orbitals. Tables of f_i as a function of d are available for every atom or ion.

The last term in eqn 1 represents the effect of thermal and crystal lattice disorder. It is known as the Debye–Waller factor, after Peter Debye who introduced it and Ivar Waller who further developed the theory. Using Bragg's law, the factor $\sin\theta/\lambda$ can be recognized as $1/2d_{hkl}$. The Debye–Waller factor is therefore a function which decreases as the resolution increases, proportionally to $\exp[-1/d^2]$. This makes it a Gaussian function of $(1/d)$. It was noted in Chapter 4 that the Fourier transform of a Gaussian shape is also a Gaussian shape. The Debye–Waller factor is usually referred to as 'the B factor'.

The Debye–Waller factor, then, smears the atomic electron density by a Gaussian shape, representing thermal disorder and crystal disorder. The quantity B represents the breadth of smearing. If the atom has a root-mean-square displacement u, then $B = 8\pi^2u^2$. Both the B factor and the atomic scattering factor f vary with the resolution d to reduce the structure amplitudes as d increases.

In work at very high resolution, it may be necessary to consider that the smearing of the atomic electron density may not be spherical. This is done by using a B factor which varies according to direction, known as an anisotropic B factor.

Using the compact notation of Box 5.1 eqn 4, the structure factor eqn (1) can be written

$$F(\boldsymbol{h}) = \sum_{N} f_i \exp[2\pi i \boldsymbol{h} \cdot \boldsymbol{x}_i] \exp[-B_i \sin^2\theta/\lambda^2]. \tag{2}$$

The summation is over all the atoms in the unit cell, which may include several copies of the basic structure, due to crystal symmetry.

A large value of B means that the diffraction fades away to nothing at relatively low θ, so that the only observable reflections have a large spacing, d, restricting the resolution that can be achieved. Only well-ordered crystals with a low B factor can be studied at high resolution. For typical protein crystals B is in the region of $20\,\text{Å}^2$. Studies beyond a resolution of $1\,\text{Å}$ are only possible if B is less than about $10\,\text{Å}^2$.

Moving-crystal techniques

At fixed wavelength, a Bragg reflection only occurs if the incident beam makes exactly the correct angle with the reflecting planes. Considering the slight disorder in a real crystal, to be sure that the reflection has been fully excited, the crystal must be rotated through its reflecting position. All fixed-wavelength observing techniques use a rotating crystal.

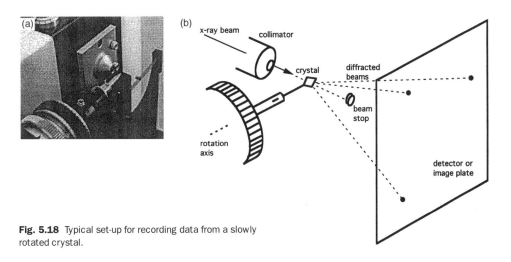

Fig. 5.18 Typical set-up for recording data from a slowly rotated crystal.

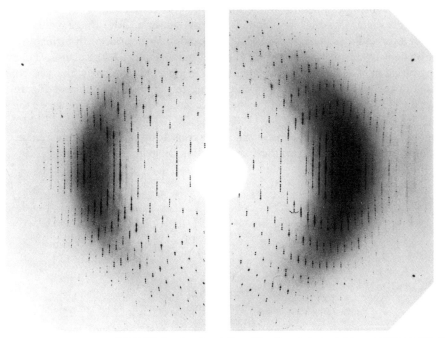

Fig. 5.19 Small-angle rotation diffraction photograph from a protein crystal. The vertical white line is the shadow of the bar supporting the beam stop.

Usually, the crystal is rotated about an axis perpendicular to the X-ray beam (Fig. 5.18). For a protein crystal, a very large number of diffracted beams may be recorded, and it is necessary to avoid them overlapping. By rotating the crystal through a small angle (typically 1–2°), only a relatively small number of reflections is recorded on the film or image-plate, and they can be observed separately. Figure 5.19 shows a crystal diffraction pattern recorded by this method, used by Desmond Bernal in the 1920s, and re-introduced for large molecules in the computer era by Uli Arndt and Alan Wonacott, 50 years later.

If a detector with continuous read-out is available (solid state or CCD detector), the overlapping problem does not arise, and a continuously rotated crystal may be observed.

For macromolecular crystals to be observed at atomic resolution, very many diffracted beams must be recorded. At the modest resolution of 2 Å, for example, a moderate-sized protein unit cell edge of 50 Å will give 25 orders of diffraction along this direction. In three dimensions, h, k, and l can all vary between about -25 and $+25$ (assuming the three cell dimensions are of similar magnitude), and (very roughly) $50 \times 50 \times 50$ or 125 000 different Bragg reflections exist to this resolution. (The number of independent reflections is reduced by a factor 2 when Friedel's law applies (Chapter 4), and the number may be further reduced by crystal symmetry, but is still substantial.) A more exact figure for the number of independent reflections is given later in this chapter.

Interpretation of a strong Bragg reflection

In one dimension, the scattering density can be analysed by Fourier methods into a series of harmonics, or terms of different frequency. If the variation of scattering density exhibits a strong third harmonic, the third-order diffraction will be strong. Figure 5.20 shows an example of a wave with a strong third-order Fourier term, giving it three prominent peaks in the repeating unit, and strong third-order diffraction.

In just the same way, each set of indices identifies a set of (hkl) Bragg planes running through a three-dimensional crystal. The corresponding Fourier term forms a wave of electron density, the peaks of which lie in planes parallel to the Bragg planes. If there is a strong variation of density on these planes (such as a layered structure with layers running parallel to the planes), the corresponding Bragg reflection will be strong. Figure 5.21 shows the (110) lattice planes, and Fig. 5.22 indicates a layered structure where the scattering density forms layers along the (110) planes.

Figure 5.22 shows layers of density which lie precisely on the Bragg planes. In this case the phase of the wave that forms the corresponding Bragg reflection is zero, since one of

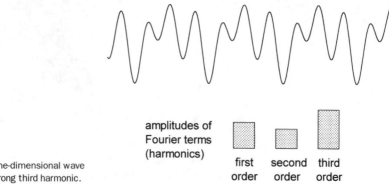

amplitudes of
Fourier terms
(harmonics)

first   second   third
order   order   order

Fig. 5.20 One-dimensional wave exhibiting strong third harmonic.

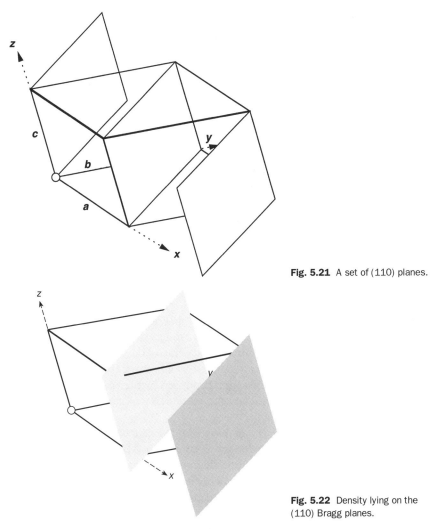

Fig. 5.21 A set of (110) planes.

Fig. 5.22 Density lying on the (110) Bragg planes.

its peaks lies on the origin of each unit cell. If the layers are offset from the Bragg planes, the phase of the reflection will not be zero; if, for example, they lie halfway between the Bragg planes as in Fig. 5.23, the phase of the reflection is 180°.

This provides a physical interpretation of the amplitude $|F_{hkl}|$ and phase ϕ_{hkl} of each Bragg reflection. The amplitude is large if the h,k,l planes correspond to large variations of electron density, and the value of the phase defines where the strongest electron density lies in relation to the h,k,l planes. Sometimes, gross features of a crystal structure can be discerned from the X-ray diffraction pattern.

Diffraction when the X-ray beam is not monochromatic

When Max von Laue's assistants Walter Friedrich and Paul Knipping took the world's first X-ray diffraction photographs in 1912, using a zinc blende crystal (Fig. 1.10), their

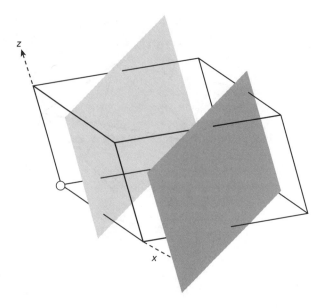

Fig. 5.23 Density lying between the (110) Bragg planes.

X-ray generator was operated under conditions that produced a wide spectrum of X-rays. Such diffraction photographs, produced from a range of wavelengths, are known as Laue photographs. The large blackened area in the centre of the photograph is caused by the direct X-ray beam, which was not intercepted by a beam-stop before it reached the film. Around it are various orders of diffraction produced by the large zinc blende crystal irradiated by a wide X-ray beam. Already the tendency for Laue photographs to generate rings of diffraction spots was visible.

When the X-rays striking the crystal have a wide range of wavelengths, the diffraction is known as *Laue diffraction*. Under these conditions, Bragg's law (eqn 5.3) can be satisfied for any incident beam direction defining θ, by an appropriate value of λ. If the spectrum of incident X-rays is wide enough, almost every set of h,k,l lattice planes can be excited simultaneously. θ still represents the glancing angle between the Bragg planes and the X-ray beam. It no longer measures the resolution associated with the reflection, because λ is variable. Planes nearly parallel to the incident beam reflect near the centre of the diffraction pattern, while those that are perpendicular are scattered so that θ approaches 90°.

Under Laue conditions it is not necessary to rotate the crystal to make sure the reflections are fully excited, because λ varies continuously. A Laue diffraction pattern obtained from a stationary protein crystal is staggeringly intricate (Fig. 5.24). It is interesting to compare the tiny diffraction spots produced by the extremely parallel X-ray beam of a synchrotron interacting with a small perfect crystal, with the cruder results of the first X-ray diffraction experiment.

Laue photographs from a very intense wide-spectrum X-ray beam, as in a synchrotron, may be made with a very short exposure (in the millisecond range), giving information about transient states. However, the technical conditions are very demanding. A high degree of perfection and stability must exist in the crystal throughout the exposure. The diffraction pattern must be recorded on a medium giving very high resolution, and orders of diffraction are often difficult to resolve from each other. It is necessary to correct for the

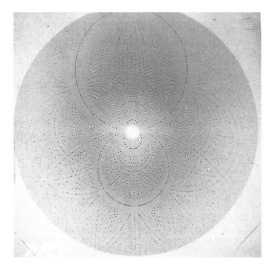

Fig. 5.24 Laue diffraction photograph of a protein.

spectrum of incident intensity at different wavelengths and the variable efficiency of detection. Another problem is that, since harmonic sets of planes such as h,k,l and $2h,2k,2l$ are parallel to each other, all scatter in precisely the same direction in a Laue photograph. The technical problem of synchronizing the transient state in every unit cell of the crystal is severe.

For these reasons, Laue methods are rarely used for primary structure determination. Increasing use will be made of them to study transient changes within a crystal whose structure is already known, if these transient changes can be triggered simultaneously in the whole crystal.

Relations between structure factors generated by symmetry

If the crystal exhibited centrosymmetry, all F_{hkl} would be real: the phases would be $0°$ or $180°$ (Chapter 4). But since crystals of a biological material are never centrosymmetric, this possibility need not be considered.

For a crystal composed of normal scattering density, Friedel's law (Chapter 4) states that the Fourier transform is centrosymmetric. In one dimension, the relation was between the scattering with index h and that with index $-h$. In three dimensions, Friedel's law requires that $|F_{hkl}| = |F_{-h,-k,-l}|$ (Box 5.3).

If the crystal structure has rotational symmetry, the Fourier transform shows the same rotational symmetry. If the crystal possesses screw symmetry, the amplitudes of the structure factors $|F_{hkl}|$ exhibit pure rotational symmetry, but the phases are changed. There are specific relationships between the phases of reflections linked by screw symmetry. Thus, rotational and screw symmetry both reduce the number of independent structure factors (Box 5.4).

As explained in Chapter 2, it is convenient to choose a crystal unit cell which fits to the lattice symmetry. With several types of crystal symmetry, this can only be done by choosing a non-primitive unit cell larger than the smallest possible or 'primitive' cell (refer to

---

Box 5.3 Friedel's law in crystal diffraction

If a crystal is composed of normal scattering density (that is, if ρ is a real quantity), Friedel's law (Box 4.5) applies. In three dimensions, Friedel's law requires that F_{hkl} is the complex conjugate of $F_{-h,-k,-l}$:

$$F_{hkl} = F^*_{-h,-k,-l}$$

which may be written:

$$F_h = F^*_{-h}. \tag{1}$$

Following the definition of the complex conjugate (Box 3.4),

$$|F_h| = |F_{-h}| \text{ and } \phi_h = -\phi^*_{-h}. \tag{2}$$

Exceptional cases will be considered later, where Friedel's law is not obeyed.

Box 5.4 Rotational symmetry and the diffraction pattern

If the crystal structure has rotational symmetry, the distribution of structure amplitudes $|F|$ shows the same rotational symmetry. For example, if there is a 2-fold axis parallel to the b axis, the 2-fold symmetry relates the reflections with indices h,k,l and $-h,k,-l$ and $F_{hkl} = F_{-h,k,-l}$. If the crystal's rotational symmetry is screw symmetry, the structure amplitudes $|F_{hkl}|$ show the same 2-fold symmetry, but the phases may differ according to specific rules. Coupled with Friedel's law, this means that the set of four $|F|$s with indices h,k,l, $-h,k,-l$, $h,-k,l$ and $-h,-k,-l$ are equal. The 2-fold rotational symmetry has halved the number of independent $|F|$s.

If you follow through the same argument for an orthorhombic crystal with three perpendicular 2-fold axes, you will find all the eight $|F|$s with indices $\pm h,\pm k,\pm l$ are equal. The four equivalent positions in the unit cell of P222 have reduced the number of independent $|F|$s by a factor 4.

The effects of rotational symmetry can be checked by calculating F_{hkl} from eqn 1, Box 5.2, including all the scattering centres generated by the appropriate symmetry, but the proof will involve some heavy algebra!

Rotational symmetry gives rise to a number of symmetry-related equivalent positions in the unit cell. It reduces the number of independent $|F|$s by the same factor.

Box 2.3). Choosing a non-primitive cell has the effect of making a fraction of the structure factors zero. These structure factors are referred to as *systematic absences* (Box 5.5).

When screw rotational symmetry is present, there are other systematic absences among reflections in lattice planes perpendicular to the screw axis. These systematic absences give evidence that screw symmetry exists.

In practice, therefore, the range of reflections that need to be observed to provide a complete data set depends upon the symmetry. Also, it is essential to make sure in calculating

Box 5.5 Diffraction by non-primitive lattices

Box 2.6 presents the space group C2 as an example where it is conventional to choose a non-primitive lattice so that its cell edges satisfy the crystal's symmetry.

It is always possible to choose a primitive lattice if required; the choice of other types of lattice is only a matter of convention, and has no effect on the physical properties. Thus the crystal diffraction could be indexed if desired on the basis of a smaller, primitive unit cell, and its diffraction properties would not be altered.

When the crystal is indexed on the basis of the larger, non-primitive unit cell, this means that some sets of indices h,k,l do not represent integral values of indices of the primitive cell. They have been introduced artificially by the conventional choice of unit cell. There can be no diffraction associated with these sets of indices, the $|F|$ values of which must therefore be considered as zero. These indices refer to reflections called *systematic absences*.

For example, in the space group C2, the C face-centring doubles the volume of the unit cell, and doubles the number of equivalent positions within it. But half of the possible h,k,l indices represent systematic absences. In C2, a systematic absence occurs if $h + k$ is odd.

The scattering from a non-primitive lattice may be calculated from eqn 1, Box 5.2, including scattering centres at all the equivalent positions in the non-primitive lattice. The proof may involve some complex algebra, but will show that certain sets of structure factors are exactly zero.

Thus, face-centring or body-centring do not affect the number of independent reflections associated with a crystal asymmetric unit.

Box 5.6 The number of independent structure factors to a given resolution

The number of independent reflections N_{ref} observable to a given resolution depends on the volume of the crystallographic asymmetric unit V_a. If symmetry creates several copies of the asymmetric unit in the cell, the unit cell is larger, creating more reflections. Due to symmetry, some of these reflections will have the same intensity $|F|^2$. These reflections are not considered to be independent.

Stating the result without proof, at a given resolution the number of independent measurable $|F|$s is given by the formula:

$$N_{ref} = \frac{2\pi V_a}{3d_{min}^3} \tag{1}$$

In this formula d_{min} is the resolution. If the resolution is doubled, say from 2 Å to 1 Å, there are eight times as many independent reflections to be measured.

It will be noted that the number N_{ref} is independent of the wavelength of the diffraction study.

the electron density from the structure factors that the structure factors used in calculation conform to the true crystal symmetry.

The number of h,k,l reflections to a given resolution depends on the size of the unit cell, increasing with the unit cell volume. However, the symmetry within the crystal unit cell imposes relationships between sets of structure factors. These reduce the number of independent structure factors, so that this number depends only on the size of the crystal asymmetric unit and not on the symmetry (Box 5.6). Whatever the symmetry, the same number of diffraction measurements provide the same amount of information for an asymmetric unit of given volume. The number of reflections to be observed increases very rapidly with resolution (as the cube of the reciprocal resolution).

Determination of crystal symmetry

Many computer programs evaluate data directly from experimental observations and automatically derive the symmetry and unit cell dimensions. These programs are not perfect, and quite often give wrong answers. In most cases, an incorrect symmetry determination will result in failure of the following steps of crystal structure determination. Sometimes the existence of an error is only revealed because ultimately the electron-density maps are uninterpretable. Very careful manual checking, and consideration of possible alternatives, is essential at the point in a structure determination where crystal symmetry is established.

How many molecules do I have in my asymmetric unit?

Protein molecules exhibit rather constant density, with a partial specific volume of about 0.76 ml/g, corresponding to a volume of 1.3 Å³/dalton. Thus, given the molecular weight of a protein, the molecular volume may predicted with accuracy of a few per cent. Protein crystals do not show this constant density, because of their solvent content. Often more than half the volume of a crystal is occupied by solvent, though the fraction is very variable. A survey by Brian Matthews (1968) showed that protein crystals, on average, have a volume of 2.4 Å³/dalton of protein, but this number may vary substantially, with a range from about 1.9 to 4.2 Å³/dalton. This Matthews volume, representing the volume of crystal per mass of protein, is extremely useful.

An estimate of the solvent content of a given protein crystal may be made by measuring its density. The density of the solvent in equilibrium with the crystal can then be used to work out what fraction of crystal volume is occupied by solvent, and what fraction by protein. Such measurements are, however, rarely made nowadays. Usually the asymmetric unit volume obtained from diffraction data suffices to give the information needed.

To take a specific example, suppose you are working with a 50 kDa protein. The volume of the protein molecule will be close to 63 000 Å³. Using the range of Matthews volumes quoted, the volume of the crystal corresponding to one molecule is probably between 95 000 Å³ and 205 000 Å³, frequently in the region of 120 000 Å³. If the asymmetric unit volume is found to be about 150 000 Å³ you can be sure there is one molecule in the asymmetric unit; if it is about 250 000 Å³ there are two. It would only be in the rare case where the volume of the asymmetric unit is around 200 000 Å³ that there would be

Table 5.1 Numbers of reflections in some protein structure determinations

Protein	Molecular weight in asymmetric unit	Resolution (Å)	Number of reflections	Year
Haemoglobin	34 000 (half-molecule)	5.5	600	1958
Myoglobin	17 000	2.0	9000	1960
Crambin	4600	0.54	112 000	2000
50S ribosomal subunit	1.6 million	2.4	666 000	2000

much uncertainty in distinguishing these situations, and a crystal density measurement would be required.

How many reflections must I measure for my protein?

Box 5.6 shows how there is a constant relationship between the number of reflections to a given resolution and the asymmetric unit volume, irrespective of symmetry and cell dimensions. To give the formula some reality, Table 5.1 shows how many independent reflections exist in a few structure determinations at a variety of resolutions.

Further reading

Most crystallography textbooks include other details in presenting crystal diffraction. The concept of reciprocal space, omitted from this account, is simply presented by (amongst many others):
Glusker, J.P. (1994). *Crystal structure analysis for chemists and biologists.* VCH, New York.

Hammond, C. (1997). *The basics of crystallography and diffraction.* International Union of Crystallography/ Oxford University Press, Oxford.

Rhodes, G. (2000). *Crystallography made crystal clear* (2nd edn). Academic Press, San Diego.

The geometry of reciprocal space is used with great advantage in the Ewald construction, which gives clear insight into the geometry of crystal X-ray diffraction patterns. The link between Bragg's idea of reflection and Ewald's method of analysis is given briefly but clearly by:
Giacovazzo, C. (1992). In *Fundamentals of crystallography* (ed. C. Giacovazzo *et al.*), pp. 154–5. International Union of Crystallography/Oxford University Press, Oxford.

Rhodes, G. (2000). *Crystallography made crystal clear* (2nd edn), pp. 50–6. Academic Press, San Diego.

The result presented in Box 5.6 has been given in slightly different forms by:
Blundell, T.L. and Johnson, L.N. (1976). *Protein crystallography*, p. 248. Academic Press, New York.

Rhodes, G. (2000). *Crystallography made crystal clear* (2nd edn), pp. 56–7 and 173. Academic Press, San Diego. *The form given here is presented in:*
Blow, D.M. (2001). Non-crystallographic symmetry. In *International tables for crystallography*, Vol. F (ed. M.G. Rossmann and E. Arnold), pp. 263–8. International Union of Crystallography/Kluwer, Dordrecht.

Part 2

Practice

6

Intensity measurements

Once a suitable crystal has been obtained, a molecular structure investigation requires measurement of the intensities of as many Bragg reflections as possible. In this chapter, some of the options that must be decided by the experimenter will be considered, and some of the criteria used to assess the accuracy and completeness of the data will be presented.

Data collection strategies

The experimenter has to make a number of strategic decisions in collecting the crystal intensity data. These include:

- What X-ray source should be used?
- What X-ray detector should be used?
- Under what conditions should the crystal be maintained?
- How long should each crystal be exposed?
- What data collection technique will be used?
- What resolution limit should be applied?

The choice of source and detector will depend largely on what is available, but the major decision is whether to use facilities in the home laboratory or whether to use a synchrotron at a central facility.

Radiation damage

The energy released by absorption of X-rays in a crystal inevitably damages it. The process of radiation damage increases crystal disorder and reduces the intensity of scattering. The experimenter will ultimately have to abandon data collection from the damaged and disordered crystal. Under ideal experimental conditions, all the useful diffraction data can be obtained from a crystal long before radiation damage takes its toll, and radiation damage does not create a practical problem. At the other end of the scale, it may be necessary to combine the measurements from many crystals in order to obtain a complete set of diffracted intensities.

There is no definite criterion to decide when a crystal is so badly damaged that it must be discarded. But if the measurements are going to be of highest quality, any observable change is bad news. The most serious effects occur in the part of the diffraction pattern at the highest observed resolution, where the observed intensities of the Bragg reflections will be altered most rapidly.

The first observable effect of radiation damage is usually a reduction of high angle intensities due to increased disorder. (The Debye–Waller factor has increased, see Chapter 5). It would be usual to stop measurements when the intensities of high-resolution measurements are significantly reduced.

It is sometimes useful to estimate the Debye–Waller factor continuously as data collection proceeds. A rough correction can be made for the increased factor, but this type of data manipulation should be avoided whenever possible.

Once it has begun, radiation damage seems to continue at a steady rate, even if the exposure ceases. This is probably due to relatively slow effects initiated by free radicals produced by X-rays in the crystalline sample. If radiation damage is a serious problem in intensity measurement, there is therefore great advantage in carrying out intensity measurement rapidly and without interruption. This has important consequences in the selection of X-ray generators and detectors.

Generators and detectors

The main types of X-ray generator have been mentioned in Chapter 1. Laboratory X-ray generators provide an accessible and reasonably reliable source. Usually a rotating-anode generator is used. Although the X-ray beams produced by these generators are very intense, they are far weaker than the beam available at a synchrotron, and observations will proceed 100 or so times more slowly. In the future, fixed-anode microfocus tubes seem likely to become useful where crystals are small.

If the crystals are very small, seriously disordered, or extremely unstable, it may be essential to accept the difficulties, delays, and pressures of using a central synchrotron facility. If none of these applies, home facilities (of suitable quality) can produce equally good results and will probably allow the experimenter to work more carefully, and to get on with the work immediately.

Some currently used detectors were also mentioned in Chapter 1. In the 1990s image plates were the most widely used: a disadvantage of image plates is that the exposure is interrupted at frequent intervals to read the data off the plate and to clear the stored image. This may use more time than the actual exposure, and throughout it crystal radiation damage progresses! To reduce the problem, some systems use more than one image plate. Detectors that allow continuous read-out of the image can avoid this problem (CCD detectors and solid-state detectors). These detectors are now used more frequently at synchrotrons, but at the moment available CCD and solid-state detectors are rather small in area—so either they record only a small part of the diffracted energy, or the detector has to be placed close to the crystal, giving a higher level of general scattered radiation which interferes with accurate intensity measurement. This matter is returned to later in the chapter.

Data may be generated by a fast CCD detector at several megabytes per second, possibly too fast for the equipment to handle, so the real rate of data collection may be limited by the power of the computer that accepts its output.

Conditions at the crystal

Concerning the conditions at the crystal, the most important decision is the temperature. In liquid nitrogen at 100 K, the effects of radiation damage are so much reduced that they are often negligible. Techniques for cooling protein crystals to liquid nitrogen temperature before exposure are now routine, and cause little damage to most crystals. Indeed, the cooled crystals are often better ordered. Crystals must be cooled very rapidly through the temperatures where ice crystals may grow, to below 190 K where water forms a stable, glassy state. A bonus from using a very cold crystal is that it does not need to be enclosed—it just sits in a cooling jet of dry nitrogen (Figs 6.1 and 6.2) (see Further reading for more detail).

But sometimes, cooling cannot be achieved without unacceptable loss of crystalline order. It may then be necessary to work with the crystal in equilibrium with mother liquor in liquid form. In these cases, maintaining the crystal just above the freezing point of its mother liquor (often substantially below 0°C) usually reduces the rate of radiation damage very significantly, compared to room temperature operation. The crystal must be enclosed in a capillary of very thin glass, which does not absorb X-rays, but prevents evaporation (Fig. 6.3).

Special precautions may be needed to maintain the crystal in the desired chemical state. Sometimes this just means making sure that pH, cofactor and ligand concentrations, and any mother liquor, are properly controlled. In other cases, a flow cell may be used to allow

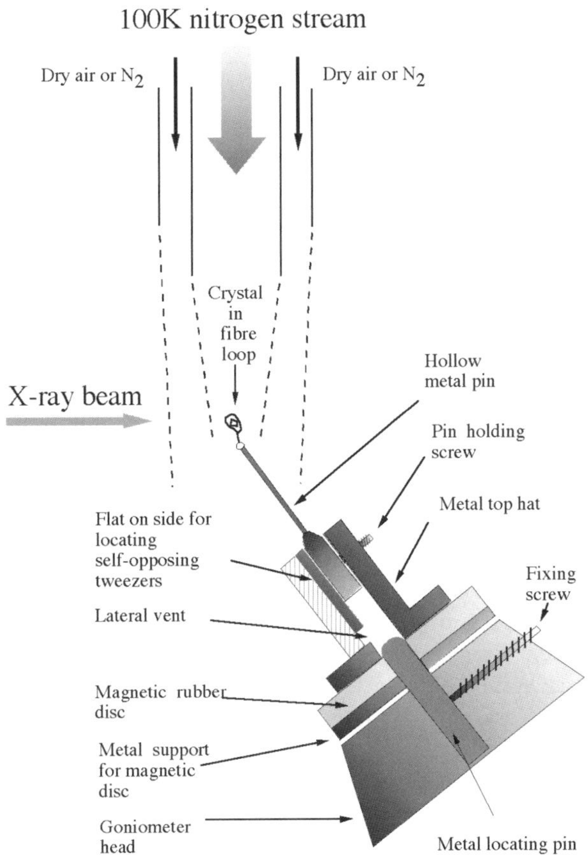

Fig. 6.1 A typical experimental arrangement for cryocrystallographic data collection (reproduced from Garman and Schneider (1997) by permission of the International Union of Crystallography).

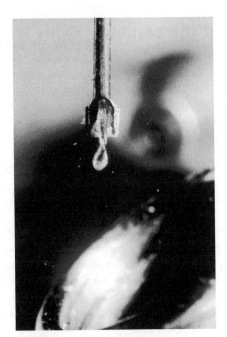

Fig. 6.2 Crystal suspended in X-ray beam on a fibre loop at 100 K (courtesy of Silvia Onesti).

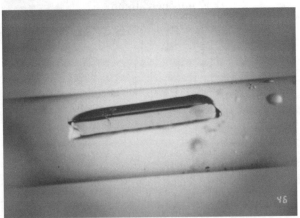

Fig. 6.3 Protein crystal in equilibrium with liquid solvent in a thin-walled capillary (courtesy of Alice Vrielink).

the same crystal to be studied under a variety of conditions, or to maintain the concentration of a substrate that is being consumed.

As described in Chapter 5, the usual configuration for intensity observation is to rotate the crystal slowly perpendicular to a monochromatic X-ray beam. In order to minimize the exposure of the crystal to X-rays, it is possible to devise strategies of crystal orientation which excite every independent Bragg reflection in the most efficient manner. Such a strategy may require careful orientation of the crystal before measurements begin, but data collection is now so rapid and versatile that this is not usually considered necessary.

At the opposite extreme, accurate orientation of some crystals would use up an unacceptably large part of their brief expected lifetime. In his work on virus crystals, Michael Rossmann introduced the 'American method' (shoot first and ask questions later), in which a series of crystals are set up in essentially random orientation, using a large enough series to obtain adequate completeness of intensity recording.

The Laue method of data collection using a 'white' X-ray spectrum from a stationary crystal appears attractive at first sight, but the practical difficulties mentioned in Chapter 5 limit the accuracy and completeness that can be achieved. However, this method is essential if data collection is to be completed in a very short time (to observe a transient state, for example).

The available resolution depends mostly on crystalline order. With a large, well-ordered crystal, an experimenter may be able to achieve his aim in structure determination without pressing for the highest resolution; but more frequently the struggle is to improve resolution. If diffraction is weak, one needs larger crystals, more intense radiation, and a larger, more sensitive detector.

Moving the detector further back from the crystal improves accuracy by reducing the background of incoherently scattered radiation from the crystal. But as the detector is moved back, it intercepts a smaller range of θ angles (Fig. 6.4) and the available resolution has been reduced. The detector may also be moved off-centre, which will extend the accessible values of θ, but with the loss of many observations at lower θ (Fig. 6.5). This strategy is rarely used

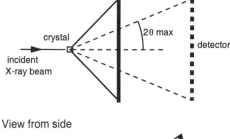

Fig. 6.4 Moving the detector back from the crystal reduces the range of θ angles that can be observed.

View from side

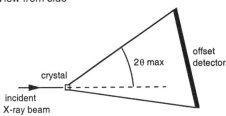

View along incident x-ray beam

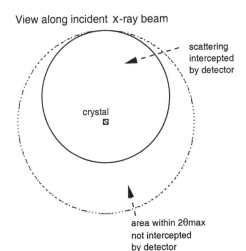

Fig. 6.5 A detector moved off-centre reaches to a larger θ angle, but intercepts only a fraction of the reflections excited within this angle.

since the detector must be placed in several different positions, needing several times as much exposure. An expert compromise may be crucial for weakly diffracting crystals.

Measures of quality of intensity data

To assess the consequences of these compromises, various measures of quality are used by crystallographers. Many of these are based on a statistic called R, which originally stood for 'reliability factor'. In essence R measures accuracy as a ratio:

$$\frac{\text{mean difference between values which should be the same}}{\text{mean magnitude of measured values}}$$

averaged over a large number of cases. If R is small, the values are accurately determined. The value of R indicates the average fractional error.

Crystal symmetry may require that the intensity of a particular observed reflection (hkl) be identical to that of some other reflection. For example, a 2-fold symmetry axis parallel to b requires that the reflections (hkl) and ($-h, k, -l$) have the same intensity. The differences between symmetry-related intensity measurements may be used to calculate an R factor called R_{sym} (Box 6.1). If intensity measurements are accurately reproducible, R_{sym} is small. Any random or systematic errors in intensity measurement will increase R_{sym}.

Many R factors, including R_{sym}, vary strongly with resolution, and it is customary to compute them for a number of small ranges of resolution (known as 'shells' or 'bins').

Box 6.1 describes how a very similar quantity, R_{merge}, may be calculated. Many authors do not distinguish R_{merge} and R_{sym}.

In the past few years, both of these R factors have been criticized for lack of statistical correctness. If reflections are measured many times, the values given by R_{merge} and R_{sym} become increased. An improved statistic, R_{meas}, has been proposed, but there are few examples of its use so far (for further discussion see Diederichs and Karplus 1997).

Under good conditions, using well-ordered crystals of proper size, small absorption, a good detector and excellent technique, R_{sym} may be as low as 0.05 (also stated as '5%'). If R_{sym} rises above 0.2 or so at high θ, this indicates a limit beyond which the data are becoming too inaccurate to be useful (Fig. 6.6).

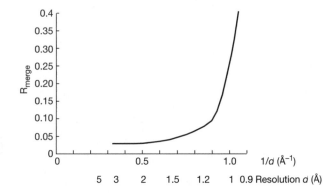

Fig. 6.6 Variation of R_{merge} with resolution for the Mn,Ca complex of concanavalin A (data courtesy of Helen Price and John Helliwell, private communication).

Box 6.1 Accuracy of the intensity measurement, R_{sym} and R_{merge}

Suppose separate intensity measurements have been made of reflection h and its n symmetry-related partners. Each measurement of a symmetry-related reflection may be identified as $I_i(h)$, where i runs from 1 to n, n being 2 or more. The best estimate of $I(h)$ is the mean of all the measurements $\bar{I}(h) = \sum I_i(h)/n$. (To be more precise, weights w_i should be applied to reduce the influence of less reliable measurements, in which case $\bar{I}(h)$ is $\sum w_i I_i(h)/\sum w_i$.)

The discrepancy between the ith observation of $I(h)$ and the best estimate is $I_i(h) - \bar{I}(h)$, and the mean discrepancy of all the observations of this intensity is $\sum_n (I_i(h) - \bar{I}(h))/n$.

In a defined group of intensity measurements, let there be N for which at least two symmetry partners have been measured. To form R_{sym} the mean discrepancy for each of these N reflections is compared to their mean intensity:

$$R_{sym} = \frac{\sum_N [\sum_n (I_i(h) - \bar{I}(h))/n]}{\sum_N \bar{I}(h)} \tag{1}$$

R_{sym} can be thought of as the mean error of an intensity measurement, compared to the mean intensity, averaged over the chosen group of reflections.

It is very important to note that R_{sym} is calculated on the basis of intensity measurements. At the end of the day, we are usually more interested in the fractional error of the structure amplitudes $|F|$, where I is proportional to $|F|^2$. Since $(1 + x)^2 = 1 + 2x + x^2$, which is about $1 + 2x$ if x is small (since x^2 is then negligible), the fractional error in I is approximately double the fractional error in $|F|$. If R_{sym} is 0.04 (representing very accurate intensity measurements), the mean fractional error in $|F|$ is about 0.02 or 2%.

It is quite usual for a particular reflection to be re-measured several times, during the collection of full intensity data. A similar R factor may be calculated which includes all the measurements of any of the symmetry-related partners. This R factor is usually calculated during the process of merging the data from all the intensity measurements, and is called R_{merge}.

The R factor is a simple quantity that has been used for many years, but it is not ideal for statistical analysis. Modern techniques make it common for a reflection to be measured a large number of times, so that each individual observation has a much larger error than the final mean intensity. R_{merge} then seriously understates the final accuracy. An improved statistic has come into use in recent years.

It is essential that all the measured data are included in calculating R. It would always be possible to improve the value by omitting very weak or poorly determined reflections, but this must not be done. Omission of weak reflections is throwing away information, and could improve R_{sym} while making the structure worse. This is one reason for insisting on a statement of data completeness, which will be discussed later in the chapter.

Resolution limit

Intensities are usually measured to a fixed limit of resolution. This may be chosen to achieve a structure of the desired quality. More often, the aim is to produce an electron-density map of the highest quality, and the resolution is pushed as far as possible.

Measurement of the intensities of diffraction from protein crystals is now such a routine matter that few details are usually published. I am grateful to John Helliwell and his colleagues for providing unpublished details of their high-resolution study of concanavalin A (Deacon *et al.* 1997).

Structure factors decrease according to the B factor (Debye–Waller factor, Chapter 5) as the reciprocal resolution ($1/d_{min}$) increases, superimposing a Gaussian factor on the diffraction intensity. A further contribution to this decrease of scattering comes from the reduction of atomic scattering factor as resolution increases (Box 5.2). The fading of intensity towards higher resolution is shown in Fig. 6.7 and is plotted as a function of reciprocal

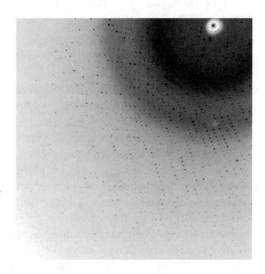

Fig. 6.7 A diffraction photograph of a Mn,Ca concanavalin A crystal, showing gradual fading of intensity towards higher resolution. The photograph extends just beyond 1.0 Å resolution. (Courtesy of John Helliwell, private communication.)

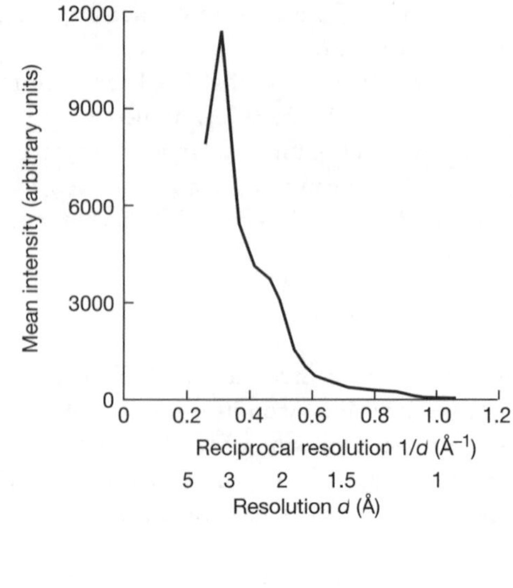

Fig. 6.8 Mean intensity for concanavalin A plotted as a function of reciprocal resolution ($1/d$). When plotted in this way, many more reflections lie at the right of the diagram ($1/d$ between 0.8 and 1 for example) than at the left (say $1/d$ between 0.2 and 0.4). (Data courtesy of Jim Raftery and John Helliwell, private communication.)

resolution $(1/d)$ in Fig. 6.8. Figure 6.8 gives a somewhat misleading impression because an area in the left-hand part of the diagram represents many fewer reflections than a similar area on the right-hand side. The same measurements are replotted in Fig. 6.9 in a way that shows equal widths of the plot for equal numbers of reflections. This emphasizes that there are many more weak reflections at the edge of the diffraction pattern, compared to the few strong ones near the centre.

Box 6.2 explains how the intensity distribution curve may be replotted as a *Wilson temperature factor plot* to give a measure of the 'overall B factor' of the data, as in Fig. 6.10, where the slope of the 'best straight line' beyond 3 Å resolution provides an estimate of $B_{overall}$. It will be seen in Chapter 12 that this is just a preliminary to assigning individual B factors to each atom or group of atoms.

Typically the overall B is around 15–20 Å² for good crystals of an average-sized protein at room temperature, tending to become larger for bigger unit cells and more flexible structures. For superbly ordered crystals, it may as low as 2.5 Å². Assuming $B = 20$ Å², crystal disorder reduces the scattered intensity to 10% of its maximum value at $d = 2.1$ Å, and to 1% of the maximum at $d = 1.5$ Å. Somewhere near this point, accurate intensity measurement becomes impossible. Figure 6.10 indicates that the concanavalin A crystals have B around 8 Å², and they are measurable to between 1.0 and 0.9 Å resolution.

There is no precise definition of the useful resolution which can be achieved from a set of intensity measurements, although in practice it is fairly 'obvious'. Some workers choose the limit at which the estimated mean intensity $<I>$ has fallen to twice the mean standard deviation of the measurements $<\sigma(I)>$. The ratio $<I>/<\sigma(I)>$ is often abbreviated to $I/\sigma(I)$. Others choose the resolution at which R_{sym} exceeds some figure, say 0.25. It is instructive to see how R_{sym} and other performance indicators change towards the edge of the resolution limit (Fig. 6.6). The extra intensity at a synchrotron may offer important improvement to the limit of measurable diffraction.

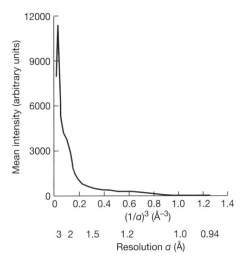

Fig. 6.9 Data from the previous diagram plotted as a function of $(1/d)^3$. When plotted in this way, equal widths on the diagram represent equal numbers of reflections. (Data courtesy of Jim Raftery and John R. Helliwell, private communication.)

Box 6.2 Resolution and the temperature factor B

The Debye–Waller factor represents the decrease of intensity in diffraction due to crystal disorder. It includes both the static disorder of the crystal and the disorder caused by thermal vibrations. The effect increases as the resolution ($1/d$) increases. According to this factor, the observed intensity I is reduced from the intensity I_0 for an ideally ordered crystal by the factor $\exp[-2B(\sin\theta/\lambda)^2]$. Since $\sin\theta/\lambda = 1/2d$ this factor may also be written as $\exp[-B/2d^2]$. The quantity B is often referred to as the 'temperature factor', though in protein crystals the most important part of it comes from intrinsic crystal disorder rather than thermal vibration.

To measure the effect of crystal disorder, it is important to allow for a second effect which reduces scattering at high resolution. Because the electron distribution around an atom is of finite size, the scattering due to a single atom is also reduced as the resolution increases. This effect causes the atomic scattering factor f to vary with resolution (Box 5.2).

The overall temperature factor, B, may be estimated using a method introduced by Arthur Wilson (1949). The intensities are divided into a series of shells according to the resolution d ($= \lambda/2\sin\theta$), and the mean intensity $\bar{I}$ is calculated in each shell. The variation of intensity with resolution is compared to the variation of the average protein f by comparing the two means $\bar{I}$ and $\bar{f^2}$. A graph is drawn of $\log_e(\bar{I}/\bar{f^2})$ against $(\sin\theta/\lambda)^2$ (Fig. 6.10). Ideally this gives a straight line with slope equal to $-B$. This graph is often referred to as a Wilson plot, but should probably be called a *Wilson temperature factor plot* to avoid confusion with another type of Wilson plot.

In practice, the presence of protein secondary structure causes characteristic departures from linearity at lower resolutions, as far as about $d = 3.0-3.5\,\text{Å}$ (around $1/d^2 = 0.1$); for larger values of $1/d^2$, the Wilson plot should be reasonably straight. The slope of this straight section measures the effective overall Debye–Waller factor B. The value of B is usually stated in units of Å^2.

For more detail see Further reading.

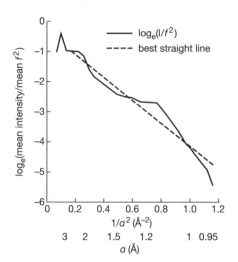

Fig. 6.10 Wilson temperature factor plot for concanavalin A (data courtesy of John Helliwell and Jim Raftery, private communication).

Completeness and multiplicity of recording

Ideally, every independent reflection within the chosen resolution should be included in an electron-density calculation. In almost any orientation of a crystal on a rotating axis, some of the required reflections are inaccessible. Very often, at least two crystal orientations are used to obtain a full coverage. On the other hand, if the crystal has high symmetry, there are many symmetry-related reflections, and a smaller range of crystal orientations may provide all the unique measurements. (The word *unique* is used here to denote the reflections needed to generate a complete set of reflections by the operation of symmetry.)

The *completeness* of a set of diffracted intensities may be defined, to a given resolution, as

$$\frac{\text{number of unique reflections measured}}{\text{total number of unique reflections}}$$

In practice, when there are tens of thousands of reflections to be measured, acceptable measurements are made for tens of thousands, but there are always a few which are omitted or deleted for one reason or another. Often the completeness of the intensity data is 95% or more; if it is much less than 95% it may indicate that some serious problem prevented complete data collection.

Symmetry-related reflections are usually more informative than repeated measurement of the same reflection, since the different geometrical conditions will give a more realistic measure of the accuracy of intensity measurement. Authors often state the total number of intensity measurements made. The factor

$$\frac{\text{total number of intensity measurements}}{\text{total number of unique reflections measured}}$$

is called the *redundancy* or *multiplicity* of intensity measurement.

Examples

These points may be illustrated by a few practical examples.

Very accurate intensity measurements (Hendrickson and Teeter 1981)

Crambin is a very small plant protein of 46 amino acids, which has spectacularly well-ordered crystals. Its structure has been determined to a resolution of $0.54\,\text{Å}$ (Jelsch *et al.* 2000). In a study of anomalous scattering by this protein, intensity measurements of the highest accuracy were required. Individual intensities were measured by diffractometer to $1.5\,\text{Å}$, using $28\,\text{s}$ of diffractometer time to measure each reflection, and various corrections were applied. The agreement between symmetry-related measurements over the whole resolution range corresponds to $R_{\text{sym}} = 0.035$. This probably still represents the 'state-of-the-art' for accurate intensity measurements.

Working with very small crystals (Kwong *et al.* 1998)

The gp120 envelope glycoprotein from HIV is a structure of potential pharmaceutical importance, but in common with other glycoproteins, large, well-ordered crystals are not available. Crystals of a complex with the CD4 receptor and an antibody gave crystalline needles less than 40 μm across, with one large unit cell dimension of almost 200 Å. For more efficient data collection at the Brookhaven synchrotron, the crystals were rotated about an axis 30° away from the long symmetry axis.

A longer X-ray exposure is required to obtain measurable diffraction from very small crystals, and radiation-damage problems were acute. Although fresh crystals diffracted beyond 2 Å, there was serious radiation damage even at 100 K.

R_{sym} and completeness values are shown in Table 6.1. For the parent data, the quality of intensity measurement has become marginal at the outer shell, at 2.5 Å resolution. For the two derivatives, measurements are seriously worse, and were limited to lower resolutions of 2.8 and 3.5 Å.

Despite the difficulty of accurate intensity measurement, the two isomorphous derivatives were able to generate a reliable structure with help from other known structures.

Working with highly disordered crystals (Mancia and Evans 1998)

When there is severe disorder in crystals, intensities are very weak at high resolution, and measurements are necessarily inaccurate. For crystals of the large protein methylmalonyl coenzyme A mutase, containing coenzyme A, the B factor was determined as 46 Å^2. For the substrate-free crystals, the B factor was much worse, 79 Å^2. Working at 100 K at the European Synchrotron Research Facility (ESRF), data were measured to 2.5–2.7 Å resolution (Table 6.2).

In order to achieve good resolution with these highly disordered crystals, measurements were continued at resolution where intensities were very weak. Even in the outermost shells the completeness of the intensity measurements was over 99%. For the outermost shell of CoA-containing crystals R_{merge} was 0.34 at 2.5 Å, and for the substrate-free crystals it was 0.38 at 2.7 Å. In these outermost shells the errors in the intensity measurements average more than one-third of the observed values.

Table 6.1 Intensity data for the HIV gp120 envelope glycoprotein complex (data from Kwong *et al.* 1998, Table 1)

Crystals	Native		K_3IrCl_6		K_2OsCl_6	
	All data	Shell comprising outer 10% of data	All data	Shell comprising outer 10% of data	All data	Shell comprising outer 10% of data
Bragg spacing limits (Å)	20–2.5		20–2.8		20–3.5	
Total measurements	113966		76739		25821	
Unique measurements	37724		28599		11982	
R_{sym}	0.093	0.247	0.115	0.202	0.143	0.182
Completeness (%)	86	62.8	90.8	82.9	72.5	62.5

Table 6.2 Intensity measurements for methylmalonyl CoA mutase (data from Mancia and Evans 1998, Table 1)

	Complex with CoA		Substrate free	
	All data	**Outer shell 2.64–2.50 Å**	**All data**	**Outer shell 2.85–2.70 Å**
Resolution (Å)	2.5		2.7	
Unique reflections	113512		44606	
B factor (Å²)	46		79	
Completeness (%)	99.6	99.7	99.2	99.2
Mean intensity/error $I/\sigma(I)$	6.0	2.2	7.8	2.0
R_{merge}	0.093	0.338	0.072	0.381

How weak were the outermost intensities? An expression in Box 6.2 allows one to calculate the effect of crystal disorder on intensity, compared to low-resolution scattering, as $\exp[-B/2d^2]$. For the coenzyme-containing crystals the factor is $\exp[-46/(2 \times 2.5^2)] = 0.025$. The resolution has been taken to the point where disorder reduced the intensities to 40 times weaker than the low-resolution data. The reduction of atomic scattering factor at this resolution makes a further contribution. Altogether, the observed intensities are about 100 times weaker.

The corresponding figure for the substrate-free crystals $\exp[-79/(2 \times 2.7^2)]$ shows that measurements were continued until they were reduced 200 times by crystal disorder, perhaps 500 times weaker in all. The authors mention that intensities for the substrate-free crystals were measured three times over, using varying exposure times to increase the dynamic range. The results show the quality of experimental technique (minimizing background scattering, high intensity, beam stability) achievable at ESRF.

Analysis of a moderate-sized protein at very high resolution (Deacon *et al.* 1997)

Concanavalin A, a plant lectin of 25 kDa, has been analysed at ever-increasing resolution since its first structure determination in 1975. A 50 mm square CCD detector was used for the structure determination at 0.94 Å at the CHESS synchrotron. A large number of intensity measurements (508 000) were merged to provide a data set of 117 000 unique reflections using four separate crystals (Table 6.3).

For measurement at large θ angle, the small detector was offset, more extremely than the example in Fig. 6.5. As a consequence, at the outer limit of resolution, the data were far from complete. At the resolution limit, the R_{merge} statistic was still respectable (0.16). It is noticeable that the low-resolution data are not particularly accurate ($R_{merge} = 0.08$), when measurements from all four crystals are included. This may be a consequence of using a very small detector which has to be placed relatively close to the crystal sample. The measurements at restricted resolution, placing the detector in a central position (crystal 2) are of better quality. R_{merge} is also increased because of the rather high multiplicity of observations (see Box 6.1).

Table 6.3 Data measurement for concanavalin A (data from Deacon *et al.* 1997)

(a) High-resolution data collection strategy

Crystal	Resolution range (Å)	R_{merge}	Data collection method
1	2.5 to 1.1	0.097	Detector offset: 30 s exposure for each 0.75° of crystal rotation
2	all to 2.0	0.059	Detector centred: quick pass: 5 s exposure for each 1.5°
3	2.5 to 0.98	0.099	Detector offset: 30 s exposure for each 0.75° of crystal rotation
4	2 to 0.94	0.104	Detector offset further: 30 s exposure for each 0.75°

X-ray wavelength was 0.92 Å.

The diffracted beams were deflected through an angle 2θ up to 59° (crystal 4).

'detector offset' means the direct beam was aligned on one corner of the square detector.

'detector offset further' means the detector was moved out of the direct beam path.

(b) Quality and completeness of data

Resolution range (Å)	No. unique reflections	R_{merge}	Completeness (%)
100.0–2.03	15647	0.081	97.7
2.03–1.61	15290	0.092	98.0
1.61–1.40	15139	0.104	97.3
1.40–1.28	14723	0.107	95.1
1.28–1.18	14429	0.112	95.3
1.18–1.11	13261	0.108	85.8
1.11–1.06	10561	0.106	68.4
1.06–1.01	8524	0.124	55.4
1.01–0.97	6551	0.161	42.6
0.97–0.94[a]	2798	0.145	18.2
Total	116923	0.093	75.4

[a] Measurements from crystal 4 only.

Very large molecular assemblies—30S and 50S ribosomal subunits (Ban *et al.* 1999, 2000; Clemons *et al.* 1999; Tocilj *et al.* 1999; Wimberly *et al.* 2000)

For the structure analysis of ribosomal subunits, very large numbers of measurements were required. Over 6 million measurements were made in obtaining 2.4 Å data for the large ribosomal subunit, merged to over 600 000 unique reflections.

Very good data collection statistics were achieved for the small subunit in 1999, using 20 cm square CCD detectors at the Brookhaven synchrotron. Measurements on the native crystals to 6.8 Å gave an excellent overall R_{sym} of 0.048. An osmium derivative diffracted satisfactorily to 5.3 Å. Another group later reported measurements to 4.5 Å, and in 2000 data were reported to 3.05 Å resolution, allowing structure determination and refinement (Table 6.4).

In 1999 the large subunit gave data at higher resolution and with a larger asymmetric unit, understandably of somewhat poorer quality. The investigators included observations to the point where the mean intensity is about twice the standard deviation of

Table 6.4 Intensity data for the 30S ribosomal subunit, molecular weight 900 000

	Data from Clemons et al. (1999)		Data from Tocilj et al. (1999)	Data from Wimberly et al. (2000)	
	'native'	Os derivative	'native'	'native'	Os derivative
Resolution	6.76	5.26	4.5[a]	3.05	3.35
Unique reflections	—	—	85991	254607	199999
Completeness (%)	95	95	93	94	98
outer shell	92	93	91	82	90
R_{sym}	0.048	0.081	0.108	0.108	0.139
outer shell	0.079	0.196	—	0.490	0.592
$<I/\sigma(I)>$				12.0	16.4
outer shell				1.8	2.4

[a] Data were measured to 3.5 Å but the outer measurements were not used.

Table 6.5 Intensity data for the 50S ribosomal subunit, molecular weight 1.6 million

	Ban et al. (1999)	Ban et al. (2000)
Resolution (Å)	120–3.75	90–2.4
Number of measurements	957850	6089802
Unique measurements	185190	665928
Completeness (%)	99	96
outer shell	—	71
R_{merge}	0.103	0.086
outer shell	—	0.691
$<I/\sigma(I)>$	13.0	25.5
outer shell	2.2	1.9

measurement. Resolution was later extended to 2.4 Å, using both the Brookhaven and the Argonne X-ray sources. The technical improvement is impressive, achieving better R_{merge} and greatly improved values of $<I/\sigma(I)>$ for both types of subunit (Table 6.5). Data are becoming very poor at the resolution limit.

Further reading

Modern intensity measurement strategy was initiated by:
Arndt, U.W. and Wonacott, A.J. (ed.) (1977). *The rotation method in crystallography.* North-Holland, Amsterdam.

Recent reviews of techniques are given in:

Carter, C.W. Jr and Sweet, R.M. (ed.) (1997). Macromolecular crystallography (Part A). *Methods Enzymol.*, **276**, 183–360. *especially:*

Dauter, Z. (1997). Data collection strategy. *Methods Enzymol.*, **276**, 326–43.

Helliwell, J.R. (1997). Overview of synchrotron radiation and macromolecular crystallography. *Methods Enzymol.*, **276**, 203-17.

Rodgers, D.W. (1997). Practical cryocrystallography. *Methods Enzymol.*, **276**, 183–202.

Other recent reviews include:

Garman, E. and Schneider, T.R. (1997). Macromolecular crystallography. *J. Appl. Cryst.*, **30**, 211–37.

Leslie, A.G.W. (2000). Data collection and reduction methods. In: *Structure and dynamics of biomolecules*, (ed. E. Fanchon *et al.*), pp. 13–35. Oxford University Press, Oxford.

For more detail on Wilson plots to estimate B factors:

Blundell, T.L. and Johnson, L.N. (1976). *Protein crystallography*, pp. 333–4. Academic Press, London.

Drenth, J. (1994). *Principles of protein x-ray crystallography*, pp. 120–2. Springer, New York.

7

Isomorphous replacement

Max Perutz has described his momentous discovery in 1953:

It ... occurred to me that the electrons of a heavy atom, being concentrated in a small sphere, would scatter in phase, and that their contribution should produce measurable intensity changes in the diffraction pattern of the protein. On the other hand, it was clear that these changes would provide the correct phases only if the attachment of the heavy atom left the structure of the protein molecules and their arrangement in the crystal unaltered. When I first tried the method I was not at all sure that these stringent demands would be fulfilled, and as I developed my first X-ray photograph of mercury hemoglobin my mood altered between sanguine hopes of immediate success and desperate forebodings of all possible causes of failure. I was jubilant when the diffraction spots appeared in exactly the same position as in the mercury-free protein, but with slightly altered intensity, just as I had hoped.

(Perutz 1992)

In this chapter, we shall see how this historic observation created the first method for determining the atomic structure of proteins, a method which is still amongst the most important.

As discussed in Chapters 1 and 3, detectors of X-rays measure the energy of light entering them. They are insensitive to the phase (Fig. 7.1). The energy of light in a diffracted beam is proportional to the square of its amplitude: all we can detect is the intensity of

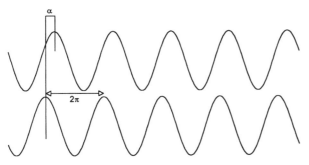

Fig. 7.1 Two waves are shown, differing in phase by α. A phase change of 2π would shift one peak of the wave to the next.

each diffracted wave. This applies to all available detectors of short-wavelength electromagnetic waves.

In order to reconstruct the electron-density distribution $\rho(x)$ from the observed scattering it is necessary to know the phase angle of the scattering associated with each Bragg reflection. The difficulty in determining the phase of a scattered wave is referred to as the *phase problem*, and the next three chapters present methods that are used to overcome it for macromolecular structure analysis.

Centrosymmetric projections

There is one situation where the phase problem is much simplified. If a structure has a centre of symmetry at the origin, all the waves that represent its Fourier transform are cosine-like (with maxima or minima at the origin), and the phases are 0° or 180°. This was presented in Chapter 4 for the one-dimensional case (see Figs 4.6 and 4.11), and applies in just the same way in two or three dimensions. All crystallographic structure determinations before 1940 dealt with centrosymmetric structures.

This does not help us directly, because biological macromolecules are never centrosymmetric, and all phase angles are possible. Having said this, there are cases where crystals of biological materials have some reflections where phase is limited to 0° or 180°. These are usually reflections with at least one index zero, depending on the symmetry: for monoclinic crystals the *h0l* reflections are always in this category. More precise conditions are defined in Box 7.1. Figure 7.2 shows how a projection of a non-centrosymmetric structure along a 2-fold axis gives a centrosymmetric projection.

Such a centrosymmetric projection requires a group of reflections to have phases of 0 or π. The condition applies to all reflections where the Bragg plane lies parallel to the projection axis. The cases where the phase problem is simplified in this way can be very useful.

Box 7.1 Centrosymmetric reflections have real structure factors

Imagine a crystal with a 2-fold axis of symmetry (say space group P2), and think of viewing its electron density along the 2-fold axis (Fig. 7.2). Projected in this direction, all the electron density is added up along the 2-fold axis. The 2-fold symmetry axis makes sure that the projection has 2-fold symmetry on the plane of projection. The two-dimensional projection has a centre of symmetry. Other kinds of even-fold symmetry (4-fold or 6-fold) will also produce a centrosymmetric projection.

It may be shown that those reflections whose reflecting plane is parallel to the even-fold symmetry direction have structure factors which are real: they have phases of 0 or π (180°). In monoclinic crystals, which have a 2-fold or 2-fold screw axis parallel to **b**, the *h0l* reflections have this property. These reflections are called *centrosymmetric reflections*. In tetragonal and hexagonal crystals which have a 4- or 6-fold axis parallel to **c**, the *hk0* reflections have real structure factors. There may also be 2-fold axes in directions perpendicular to **c**, generating more centrosymmetric reflections.

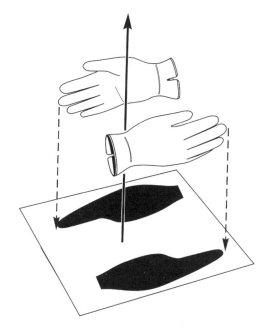

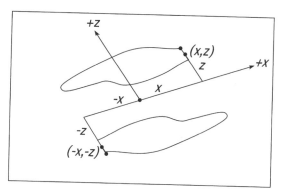

Fig. 7.2 A pair of non-centrosymmetric objects, arranged to have 2-fold symmetry. When projected parallel to the 2-fold symmetry direction, the projected density forms a centrosymmetric arrangement in two dimensions. (Drawn by Neil Powell.)

The isomorphous replacement method

The isomorphous replacement method is based on the comparison of two crystals, which differ because one or more strong scattering centres exist in one crystal which are absent in the other (Fig. 7.3). In the most usual case, a heavy metal ion or heavy metal compound binds tightly to one site (or a few sites) in the crystal asymmetric unit of a protein, which can equally well be crystallized in the absence of the heavy metal. Max Perutz achieved this by binding a mercury compound to a sulphydryl site on the surface of the haemoglobin molecule.

Since, in the majority of cases, a protein is first crystallized in the absence of heavy atoms, the crystal without the strong scattering centres may be referred to as the *parent*, and a crystal including the heavy atoms may be called the *derivative*. (Sometimes the

Fig. 7.3 A pair of isomorphous crystal structures, differing at a single site.

parent crystal form is referred to as 'native'.) A full structure determination by isomorphous replacement requires several derivatives of the same parent.

The word 'isomorphous' refers to the need to have a parent and derivative crystals with the same unit cell dimensions and symmetry. In the first applications of the method (in simple organic compounds) one atom was replaced by another: a chloride might be replaced by iodide, for example. In the majority of cases nowadays, the parent is native protein without heavy atoms, and in the derivative a heavy atom is introduced by chemical reaction or by soaking the crystals with a ligand which binds to specific sites. These changes always disturb the structure, so isomorphism is never perfect. The inserted group will inevitably distort the surrounding protein structure, and will probably force some solvent out of the derivative crystal. These changes and distortions are referred to as *non-isomorphism*.

The isomorphous replacement method depends on a simple idea (Box 7.2):

> *the scattering by a derivative crystal,* PH, *containing heavy atoms is the sum of the scattering by the parent crystal,* P, *and the scattering by the added heavy atoms,* H.

A derivative may be called isomorphous with a parent if this statement is obeyed with sufficient accuracy. The accuracy needed depends, of course, on how the isomorphous replacement method is to be used. The necessary accuracy gets harder to achieve as the resolution is increased.

Box 7.2 The isomorphous replacement equation

Using Box 5.2 eqn 2, the scattering factor of the parent crystal, P, is:

$$F_P(\boldsymbol{h}) = \sum_N f_i \exp[2\pi i \boldsymbol{h} \cdot \boldsymbol{x}_i] \exp[-B_i \sin^2\theta/\lambda^2].$$

Now suppose that the derivative contains one heavy atom, H, in addition to the N atoms of P. Then the scattering factor of the heavy atom derivative, PH, would be obtained by adding the scattering of the heavy atom into the summation:

$$F_{PH}(\boldsymbol{h}) = \sum_N f_i \exp[2\pi i \boldsymbol{h} \cdot \boldsymbol{x}_i] \exp[-B_i \sin^2\theta/\lambda^2]$$
$$+ f_H \exp[2\pi i \boldsymbol{h} \cdot \boldsymbol{x}_H] \exp[-B_H \sin^2\theta/\lambda^2]. \tag{1}$$

In practice, the heavy atom derivative may contain more than one atom, and its scattering may be expressed by a summation over several atoms, so that

$$F_{PH}(\boldsymbol{h}) = \sum_N f_i \exp[2\pi i \boldsymbol{h} \cdot \boldsymbol{x}_i] \exp[-B_i \sin^2\theta/\lambda^2]$$
$$+ \sum_M f_{H,i} \exp[2\pi i \boldsymbol{h} \cdot \boldsymbol{x}_{H,i}] \exp[-B_{H,i} \sin^2\theta/\lambda^2], \tag{2}$$

where the second summation is taken over M heavy atoms. This second term represents the total scattering F_H due to the heavy atoms. The equation can be re-written:

$$F_{PH}(\boldsymbol{h}) = F_P(\boldsymbol{h}) + F_H(\boldsymbol{h}). \tag{3}$$

Eqn 3 will be called *the isomorphous replacement equation*. It is illustrated graphically in Fig. 7.5 for the centrosymmetric case and Fig. 7.6 for the non-centrosymmetric case.

The method depends on being able to calculate the scattering by the heavy atoms H, and therefore requires a simple pattern of substitution. In the simple case represented in Fig. 7.3, just one heavy atom is introduced into the asymmetric unit at a unique site.

Sign determination for centrosymmetric reflections by isomorphous replacement

Even though a mercury atom has only about 0.2% of the electrons in the haemoglobin derivative, these electrons are concentrated into one atom. They make changes of a few per cent to the diffracted intensities.

In isomorphous replacement, each Bragg reflection has to be analysed to determine its phase. In what follows, F_P, F_H and F_{PH} will refer to one particular Bragg reflection. In the end, every observed reflection will have to be analysed in the same way.

We start with a centrosymmetric case. Horse haemoglobin has 2-fold symmetry with pairs of molecules related as in Fig. 7.2, and the $(h0l)$ reflections define a centrosymmetric projection of the structure. These reflections have phases of $0°$ or $180°$, which are referred to as $+$ or $-$, as depicted in Fig. 7.4.

If F_H can be calculated, its sign is known. Its magnitude $|F_H|$ can be assumed to be small compared to $|F_P|$ and $|F_{PH}|$, but it makes a measurable difference to

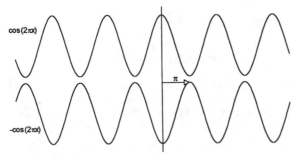

$\cos(2\pi x)$

π

$-\cos(2\pi x)$

Fig. 7.4 Both these waves are centrosymmetric about the origin. The lower wave can be seen as the negative of the other, or as a wave having a phase that has changed by π.

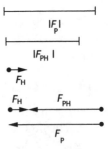

$|F_P|$

$|F_{PH}|$

F_H

F_H F_{PH}

F_P

centrosymmetric case

Fig. 7.5 The magnitudes of F_P and F_{PH} are observed, and a value for F_H can be calculated. In the centrosymmetric case, all phases must be 0 or π. The sum of F_P and F_H must equal F_{PH}, which is only achieved in one way. The sign of F_P must be negative.

Box 7.3 Determining the signs of centrosymmetric reflections by isomorphous replacement

For large molecules, in practice the heavy atom scattering is much weaker than the scattering from the rest of the molecule, so in the centrosymmetric case F_P and F_{PH} almost always have the same sign.

Since $F_H = F_{PH} - F_P$, and F_H can be calculated, the signs of F_P and F_{PH} can be decided. Figure 7.5 shows an example.

A rule can be stated. If $|F_{PH}|$ is greater than $|F_P|$, they both have the same sign as F_H; if $|F_{PH}|$ is less than $|F_P|$, they both have the opposite sign to F_H. Exceptions may arise if $|F_P|$ or $|F_{PH}|$ is less than $|F_H|$, when their signs may differ.

---

them.[1] Because $|F_H|$ is small, the signs of F_P and F_{PH} are nearly always the same. By observing whether the scattering amplitude is increased or decreased by the heavy atom, the sign of F_P and F_{PH} may be discovered (Fig. 7.5 and Box 7.3).

In this way, the signs of all the $h0l$ reflections can be determined. But what about the other reflections?

---

[1] If a variable is enclosed between vertical bars (| |, known as modulus signs), the quantity defined is its magnitude, which is always positive. For a centrosymmetric reflection whose structure factor F is negative, $|F|$ is the positive quantity $-F$. If F is -3, $|F|$ is 3. For non-centrosymmetric reflections, F is a complex number and $|F|$ is its magnitude, its length if represented as a vector on an Argand diagram, or the amplitude if it represents a wave. The symbol $|F|$ is usually read as 'mod F'.

Determination of phase angles by isomorphous replacement requires more than one derivative

Except for the special cases of centrosymmetric reflection, the phase angle may take any value between 0 and 2π (360°). The various structure factors F_P, F_{PH} and F_H can be represented as vectors with lengths corresponding to their magnitudes, and in directions corresponding to their phases. The isomorphous replacement method assumes that F_P and F_H add together to give F_{PH}, as in Fig. 7.6.

Assuming that F_H can be calculated, while for F_P and F_{PH} only the magnitudes can be observed, there are two ways of satisfying the isomorphous replacement equation. There are two different ways (shown in Fig. 7.7) to construct triangles with two sides whose lengths are $|F_P|$ and $|F_{PH}|$, on the line representing F_H. Although it provides a great deal of information, a single isomorphous derivative cannot determine a phase uniquely.

However, if more than one isomorphous derivative is available, the problem does appear to be solved. Figure 7.8 shows how a second heavy atom derivative also suggests two possible phase angles for F_P, so that the one in common satisfies both derivatives.

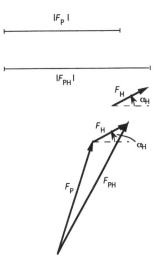

Fig. 7.6 A graphic representation of the isomorphous replacement method for phase determination. Non-centrosymmetric structure factors F_P and F_H add to F_{PH}.

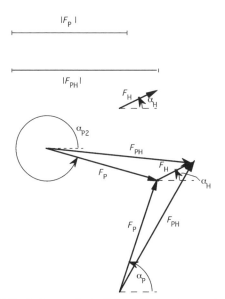

Fig. 7.7 If only the magnitudes $|F_P|$ and $|F_{PH}|$ are known, there are two ways to fit them to the calculated heavy atom structure factor F_H, giving two different possibilities for the phase α_P.

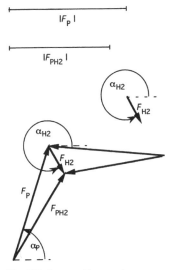

Fig. 7.8 A second isomorphous derivative also indicates two possible phases for F_P, one of which is consistent with a phase suggested by the first derivative.

When isomorphous replacement is to be used on its own to solve a non-centrosymmetric crystal structure, more than one isomorphous derivative is needed. In practice, at least three derivatives are desirable. The method is often referred to as multiple isomorphous replacement, or MIR.

Evaluating possible isomorphous derivatives

Attempts to prepare heavy-atom derivatives of protein crystals require the protein to be exposed to a heavy-atom-containing reagent. Either the crystals are soaked in the reagent or, alternatively, the protein is exposed to heavy atoms first, and then crystallized.

If crystals are exposed to the heavy atom in solution, the first requirement is to test whether the crystals remain well-ordered and have a similar unit cell under the altered circumstances. Similarly, if crystals are grown after exposure to a heavy atom reagent, the first test is that crystals can be grown which are sufficiently well-ordered to give good diffraction measurements, without serious change of cell dimensions. (Changes more than about 1% are 'serious'.)

Crystals of heavy-atom complexes are usually less well-ordered than crystals of the parent protein, but very often they give diffraction data that may be measured to a useful resolution.

Given well-diffracting and apparently isomorphous crystals, the first task is to evaluate the magnitude of the changes to the structure factors. One calculates an R factor (variously called R_{deriv}, R_{diff}, or R_{iso}) which evaluates the amount by which the structure amplitudes have been altered, as a fraction of the mean structure amplitude (Box 7.4). In practice, if R_{deriv} is less than about 0.05, the chances are that no heavy-atom site is

---

Box 7.4 Fractional structure factor change, R_{deriv}

Suppose the structure amplitude of reflection h has been measured for the parent crystal P and for the derivative PH. The magnitude of the difference between these measurements is given by $|F_{PH}| - |F_P|$ if $|F_{PH}|$ is the larger, or $|F_P| - |F_{PH}|$ if $|F_{PH}|$ is the smaller. Both these possibilities can be included by writing the difference as the modulus of $|F_P| - |F_{PH}|$, namely $||F_{PH}| - |F_P||$.

Let there be N independent reflections for which both $|F_{PH}|$ and $|F_P|$ have been measured. To form R_{deriv}, the mean difference for these N reflections is compared to the mean parent structure amplitude for the same reflections:

$$R_{deriv} = \frac{\Sigma_N ||F_{PH}| - |F_P||}{\Sigma_N |F_P|}. \tag{1}$$

R_{deriv} can be thought of as the mean fraction by which the structure amplitude is changed in forming the derivative, averaged over the chosen group of reflections.

It is important to note that R_{deriv} is calculated on the basis of structure amplitudes. R_{sym} and R_{merge}, on the other hand, are calculated from observed intensities (Chapter 6). If R_{deriv} were calculated from two sets of measurements of the same crystal, it would be about half R_{merge}.

significantly occupied. For a moderate-sized protein with one fully occupied heavy atom site, R_{deriv} between 10 and 3.5 Å resolution is usually 0.1–0.2. If the changes are larger than this, either there are several sites of substitution, or a more considerable change has been made to the crystals. A protein substituted at many sites may be a very useful isomorphous derivative, provided the sites can be satisfactorily analysed, so R_{deriv} being above 0.20 may not be bad news. However, large changes may be due to other modifications to the crystal structure. If it is greater than 0.28 or so, the chances of being able to analyse the derivative satisfactorily are dwindling.

Obviously, the above values give only general guidance. The size of the protein, the atomic weight of the heavy atoms, the accuracy of intensity measurement, and the attainable resolution all affect the situation.

Preliminary evaluation of derivatives

The preparation and evaluation of heavy atom derivatives has become such a routine matter that details are rarely published. Some examples are taken from very old publications. For some of the more recent work, I specially thank those who have allowed me to use data from unpublished sources, such as doctoral theses.

Straightforward case: trypsin inhibitor from *Erythrina caffra* (Onesti *et al.* 1991)

Hexagonal crystals of a Kunitz-type trypsin inhibitor from *Erythrina caffra* (19 kDa) diffract to 2.5 Å resolution. Isomorphous derivatives containing Pb, Ir, and two derivatives with mercury (PCMB and HgAc) gave satisfactory intensity data to resolutions between 3.0 and 2.5 Å resolution, with values of R_{deriv} that show significant changes (Table 7.1).

Weak substitution: D-glycerate dehydrogenase (Goldberg 1993)

For triclinic crystals of D-glycerate dehydrogenase (a dimer of 38 kDa units) diffraction data were collected to 3.0 Å resolution for seven putative heavy atom derivatives, all worse ordered than the parent. In Fig. 7.9, R_{deriv} is plotted against resolution for each derivative. For three of these, R_{deriv} at low resolution was in the range 0.15–0.18, while low values for the other four suggest weak substitution of heavy atoms in these four derivatives, which ultimately gave little phase information.

Table 7.1 Data for heavy atom derivatives of *Erythrina* trypsin inhibitor (Onesti *et al.* 1991)

	Resolution (Å)	R_{merge}	R_{deriv}
Parent	2.5	0.065	
Pb	2.5	0.058	0.201
Ir	2.7	0.072	0.150
PCMB	3.2	0.062	0.168
HgAc	3.2	0.062	0.216

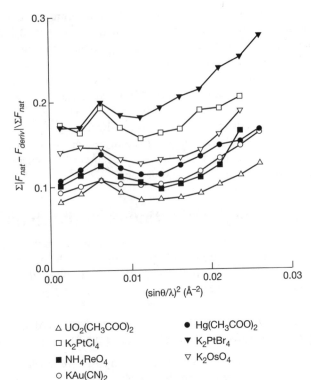

Fig. 7.9 R_{deriv} as a function of resolution for seven putative derivatives of D-glycerate dehydrogenase (reproduced by permission from Goldberg 1993).

△ UO$_2$(CH$_3$COO)$_2$ ● Hg(CH$_3$COO)$_2$
□ K$_2$PtCl$_4$ ▼ K$_2$PtBr$_4$
■ NH$_4$ReO$_4$ ▽ K$_2$OsO$_4$
○ KAu(CN)$_2$

How can the first heavy atom positions be found?

Application of the isomorphous replacement method requires a calculated value for each F_H. To see how these may be derived, we again start with the centrosymmetric case, where every phase angle can be expressed by a sign, $+$ or $-$.

Centrosymmetric difference Patterson functions

As already mentioned, in the vast majority of centrosymmetric reflections F_P and F_{PH} have the same sign, because $|F_H|$ is much smaller than they are. In every such case, the difference of $|F_P|$ and $|F_{PH}|$ gives the magnitude $|F_H|$ (Fig. 7.5), but not its sign. If we know all the $|F_H|$, how can this help to find the heavy atom positions?

Back in Chapter 4, the Patterson function was described, which may be used to determine simpler molecular structures directly from the scattered intensities. It is a Fourier summation with coefficients $|F|^2$. The result of the Fourier summation is the convolution of the structure with itself, giving the distribution of vectors between the atoms.

Just as the Patterson function could be used to determine simple molecular structures, so a Fourier summation with coefficients $|F_H|^2$ would provide the Patterson function of the heavy atom scattering distribution. In the centrosymmetric case, where F_P and F_{PH} almost always have the same sign, $|F_H|$ may be estimated as the difference between $|F_P|$ and $|F_{PH}|$. The Patterson function, the coefficients of which are $|F_H|^2$, can be calculated without knowing the signs of F_P. Deconvolution (or 'solving') of this Patterson function can generate a set of heavy atom positions.

(a)

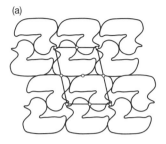

(b)

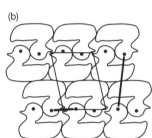

Fig. 7.10 (a) Two-dimensional lattice with 2-fold symmetry. A unit cell is shown. (b) Isomorphous lattice with one heavy atom added to every asymmetric unit. Vectors between two pairs of heavy atoms are emphasized.

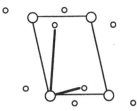

Fig. 7.11 Patterson function of the heavy atoms. Two atoms in the unit cell give double-weight peaks at the origin, two single-weight peaks within the cell.

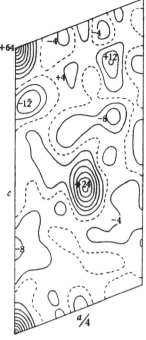

Fig. 7.12 Centrosymmetric difference Patterson projection for horse haemoglobin at 6 Å resolution. The Patterson density is shown as a contour map. The origin peak is 64 units, and the next highest peak of 24 units is double the height of any other peak. (Reproduced from Green *et al.* (1954) by permission of the Royal Society.)

Figure 7.10 is a two-dimensional example. It shows a parent structure and an isomorphous derivative projected along a 2-fold axis which gives a centrosymmetric projection. Figure 7.11 shows the corresponding difference Patterson function.

Historic example: horse haemoglobin (Green *et al.* 1954)

Horse haemoglobin is monoclinic, and has a centrosymmetric projection along the 2-fold axis. The first isomorphous protein crystal derivatives exploited contained mercury and silver bound to a protein sulphydryl group. The difference Patterson projections along the 2-fold axis have unambiguous peaks at $u = 0.136$, $w = 0.580$ (Fig. 7.12). These peaks represent the vectors between one heavy atom site and its 2-fold related site (as shown in Fig. 7.10), leading to the conclusion that there is a heavy atom site at $x = 0.068$, $z = 0.290$ (relative to the origin halfway between them, on the 2-fold axis).

This was the first difference Patterson function used in isomorphous replacement, based on only 70 reflections (since the authors cautiously rejected many uncertain measurements amongst the 128 *h0l* reflections to 6 Å resolution). Since the origin along *y* may be chosen arbitrarily, the heavy atom coordinate *y* was chosen to be 0. A single centrosymmetric projection, in this simple case, led directly to the assignment of all three heavy atom coordinates.

Non-centrosymmetric difference Patterson functions

For a non-centrosymmetric structure, the difference of $|F_P|$ and $|F_{PH}|$ no longer gives $|F_H|$ (see Fig. 7.6). However, the difference should never exceed $|F_H|$, and if it is large, $|F_H|$ is also large. It is found that a *difference Patterson* function using the Fourier coefficients $(|F_{PH}| - |F_P|)^2$ provides a slightly noisy version of the Patterson function of the heavy atoms. This Patterson function may be calculated in three dimensions. Further details are given in Box 7.5, and an example is given below.

Three-dimensional difference Patterson: cholesterol oxidase (Vrielink 1989)

Cholesterol oxidase crystallizes in space group $P2_12_12_1$. If there is a single heavy atom site in the crystal asymmetric unit, the 2-fold screw axes cause the other sites in the unit cell

---

Box 7.5 How to use a difference Patterson: simple cases

The very simplest distribution of heavy atoms would be in space group P1, the space group that has no symmetry except the lattice translations (Box 2.4). A single heavy atom substitution places one atom at the same point in every unit cell (Fig. 7.3). The only vectors between heavy atoms are lattice translations, and the difference Patterson function has density only at the origin of every unit cell. Such a Patterson indicates that there is only one heavy atom site. In P1, there is no symmetry to set the origin position. You may as well choose the origin to be the heavy atom site $(0, 0, 0)$. All the F_H have phase $0°$ when this origin is used.

A single heavy atom site in space group P2, which has 2-fold axes running along $\boldsymbol{b}$, the positions of which define an origin for x and z, is a more usual example (Fig. 2.37 and Box 2.5). The origin must be chosen on the 2-fold axes, but the choice of origin of y is arbitrary, and can be made so that a heavy atom's y coordinate is 0. The 2-fold symmetry creates pairs of heavy atom sites, so if there is one heavy atom at $(x, 0, z)$ there is a 2-fold related site at $(-x, 0, -z)$. (Figure 7.10 shows this in 2 dimensions.) The vectors between these two sites are $\pm(2x, 0, 2z)$. In the Patterson function (Fig. 7.11), there are peaks at the origin which are double-weight peaks, because they relate two atoms in every unit cell. There are also single-weight peaks at $\pm(2x, 0, 2z)$. If these are the only features in the Patterson, it tells you that there is just one heavy atom site. It is easy to work back from their coordinates to the heavy atom coordinates $(x, 0, z)$.

Next, suppose there are two heavy atom sites in space group P1 (Fig. 7.13). There will now be difference Patterson peaks representing the vectors between the two sites, at $(x_1 - x_2, y_1 - y_2, z_1 - z_2)$ and at $(x_2 - x_1, y_2 - y_1, z_2 - z_1)$, with double-weight peaks at the origin because there are two sites. The Patterson function is shown in Fig. 7.14. If the first atom is chosen as the origin, so that $x_1 = y_1 = z_1 = 0$, the coordinates x_2, y_2, z_2 of the second will be the same as one of the non-origin peaks of the Patterson.

These examples show how coordinates of heavy atoms may be deduced from Patterson functions, and emphasize that the experimenter may have to make a choice when defining the origin.

Fig. 7.13 Structure with two heavy atom sites in a simple lattice. The vectors between two pairs of atoms are emphasized.

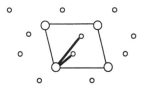

Fig. 7.14 Corresponding difference Patterson function. Two sites per unit cell give double weight at origin, and two single-weight vectors.

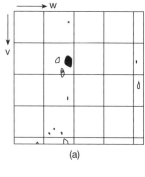

(a)

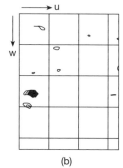

(b)

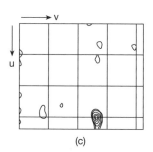

(c)

Fig. 7.15 Sections through the three-dimensional difference Patterson for a mercury derivative of cholesterol oxidase. (a) Section $u = 1/2$, (b) section $v = 1/2$, (c) section $w = 1/2$. (Reproduced by permission from Vrielink 1989.)

Fig. 7.16 With a single isomorphous replacement, two phases are equally possible. The best estimate is made by taking an average of the two possible vectors. This represents the component parallel to F_H and ignores the unknown perpendicular component.

to be separated from this site by exactly half the cell dimension along x, y, or z. Thus there are peaks in the difference Patterson on the planes $u = 1/2$, $v = 1/2$ and $w = 1/2$. Figure 7.15 shows these planes of the difference Patterson function for a mercury derivative. The major peak on the $u = 1/2$ plane indicates y and z coordinates for the heavy atom sites. The other two planes indicate x and z, and x and y coordinates, respectively. All the indications are consistent with a single heavy atom site.

Once heavy atom positions have been found for one derivative, and the amount of scattering at each site has been estimated, a rough approximate value for the phase of F_H can be calculated. Using the observed magnitude $|F_H|$, a first approximate phase angle for the protein structure factor F_P can be taken as the mid-point of the two possible values, as shown in Fig. 7.16. Even this crude information is important to help interpret other derivatives.

Electron-density difference maps

Suppose a very approximate phase indication, α_{P1}, is given by a first derivative. There is a second derivative, for which the structure amplitudes $|F_{PH2}|$ are measured. We know $|F_P|$. These allow us to make an approximate estimate of the heavy atom structure factors for the second derivative, F_{H2}, as shown in Fig. 7.17.

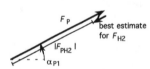

A Fourier series calculated with these approximate F_{H2} as coefficients then provides an estimate of the electron density of the heavy atom(s) included in this second derivative. Typically, these maps are calculated in three dimensions, but Fig. 7.18 shows three two-dimensional projections. The map gives a direct picture of the electron density, which is far easier to interpret than the Patterson function. Although it is inaccurate because the estimated phases are inaccurate, its peaks are usually clearer than those of the difference Patterson function. Moreover, it is consistent with whatever choice of origin (and handedness, see below) was made in interpretation of the first derivative.

Very crude approximations are made in assigning the phases to calculate these maps. How can they possibly work? The reason is that many reflections, typically tens of thousands, are used to prepare a map which shows the positions of a small number of atoms. The total errors, although huge, cancel each other out very efficiently. The 'correct' components, although individually small, are added thousands of times to give a clear signal.

A map of this kind is an electron-density difference map (often called a difference Fourier) (Box 7.6). You can use any approximate set of phases to estimate the electron-density difference between a derivative and parent. Once heavy atom sites in the second derivative have been placed, this derivative can be used to determine a new set of phases, α_{P2}, in exactly the same way. These α_{P2} phases may then be used to calculate an electron-density difference map for the first derivative. Substitutions at minor sites, which may have been overlooked in initial interpretation of the first difference Patterson function, may now be obvious. In this way, the interpretations of heavy atom sites, and estimates of F_H, are improved and refined.

Fig. 7.17 An estimate of F_{H2} is formed by the approximation that F_P and F_{PH2} have the same phase.

Of course, if you use your α_{P1} phases to calculate a difference map for the first derivative, it will give you a beautiful map showing large peaks at your originally identified sites! It is the equivalent of proving something by assuming it to be true.

Once it is possible to use two or more derivatives, phasing becomes much more accurate and difference maps for other derivatives should become much clearer. Usually

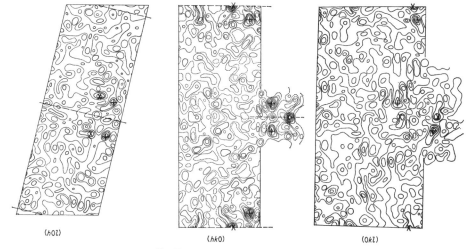

(h0l)   (hk0)   (0kl)

Fig. 7.18 Three electron-density difference projections using phases generated from a single isomorphous derivative (a chloroplatinite derivative of α-chymotrypsin). One projection (h0l) is centrosymmetric, while the other two are non-centrosymmetric. The peaks represent the positions of iodine atoms in an iodinated chymotrypsin inhibitor. (Reproduced by permission from the *Journal of Molecular Biology*, Sigler *et al.* (1966).)

Box 7.6 The electron-density difference map: interpreting later isomorphous derivatives is simpler!

An electron-density difference map estimates the difference of electron density between two isomorphous derivatives, when only one set of phases is known. In order to do this, it assumes that this set of phase angles applies to both derivatives.

For example,

$$\rho_{P}(x,y,z) = \frac{1}{V}\sum_{hkl}F_{P}\exp[-2\pi i\mathbf{h}\cdot\mathbf{x}]$$

and

$$\rho_{PH}(x,y,z) = \frac{1}{V}\sum_{hkl}F_{PH}\exp[-2\pi i\mathbf{h}\cdot\mathbf{x}],$$

so the electron-density difference is

$$\Delta\rho(x,y,z) = \rho_{PH}(x,y,z) - \rho_{P}(x,y,z)$$

$$= \frac{1}{V}\sum_{hkl}(F_{PH}-F_{P})\exp[-2\pi i\mathbf{h}\cdot\mathbf{x}].$$

The phase of F_{PH} is not known, so we make the approximation that F_P and F_{PH} have the same phase,

$$F_{H} = F_{PH}-F_{P} \text{ is approximately } (|F_{PH}| - |F_{P}|)\exp[i\alpha_{P}]$$

(Fig. 7.17). This approximation, together with an estimate of α_P, makes it possible to calculate an approximate difference density:

$$\Delta\rho(x,y,z) = \frac{1}{V}\sum_{hkl}(|F_{PH}| - |F_{P}|)\exp[i\alpha_{P}]\exp[-2\pi i\mathbf{h}\cdot\mathbf{x}].$$

at least three isomorphous derivatives are needed; additional derivatives will continue to improve the experimentally determined phase angles, using methods discussed below.

Before experimentally derived phases are used to calculate an electron-density map, it is important to improve the estimates of F_H as far as possible, using all available information. For each derivative, an electron-density difference map may be calculated (from 'unbiased' phases, derived without using data from the particular derivative, but using all other available phase information). These maps may reveal unrecognized minor sites of substitution, or other errors in placement of heavy atoms. Computational refinement procedures improve estimates of the scattering power, coordinates, and temperature factor associated with each heavy atom site. These processes are nowadays performed automatically by some software.

Handedness

In interpreting a structure by isomorphous replacement, at one point in generating the coordinates of the heavy atoms of different derivatives, it will be necessary to make an arbitrary decision. You are assuming one form for the overall arrangement of all the heavy atoms, when it would be equally possible to choose a form which is the mirror image of this assumption. The Patterson function itself is bound to be centrosymmetric, and can give no clue which choice is correct. A more detailed example is given in Box 7.7.

The consequence of this choice emerges when an electron-density map is finally calculated for the protein. This map may show L-amino acids and right-handed α-helices, in which case the 'correct' choice was made. If there are D-amino acids and left-handed helices, the choice has to be reversed.

Box 7.7 Arbitrary choice of handedness

In calculating phase angles by the isomorphous replacement method, an arbitrary choice always has to be made at some point, between a constellation of heavy atoms, or their mirror image.

The choice often arises at the time the second derivative's coordinates are assigned. As an example, follow the case of monoclinic symmetry shown in Figs 7.10 and 7.11, corresponding to the horse haemoglobin example, Fig. 7.12. If the first isomorphous derivative has a single heavy atom site, the site may be assigned an arbitrary y coordinate of 0.0. There is a centrosymmetric array of heavy atoms, and the consequence is that all the calculated F_H have phases of 0° or 180°. The method illustrated in Fig. 7.5 may be used to obtain signs as approximate estimates of the phases of F_P. At this stage all F_P are assigned phases of 0° or 180°.

When an electron-density difference map is calculated for a second isomorphous derivative, the calculated difference map will be centrosymmetric. This is a necessary consequence of the assignment of all F_P phases as 0° or 180°. If a heavy atom is really at x_2, y_2, z_2 in the second derivative, there will be a peak in the map at this position, but because the map is centrosymmetric there will be an identical peak at $-x_2, -y_2, -z_2$. The experimenter must make an arbitrary choice which of the two peaks to select as the heavy atom position.

Phases generated by multiple isomorphous replacement

Figures 7.7 and 7.8 suggest that each isomorphous derivative will offer a choice of two phase angles, one of which should agree with a phase angle offered by each of the other isomorphous derivatives. To generate Fig. 7.8 we can proceed as shown in Figs 7.19 and 7.20.

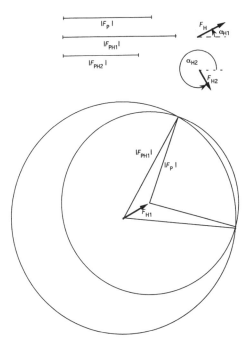

Fig. 7.19 Harker construction for a single derivative, giving two possibilities for F_P and F_{PH1}.

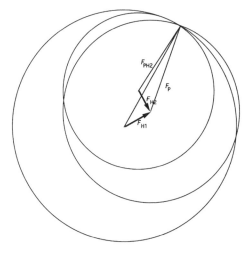

Fig. 7.20 A second isomorphous derivative identifies a single phase for F_P.

1. Draw a circle of radius $|F_P|$. This represents the possible values of the complex quantity F_P for all values of its phase angle.
2. Draw a vector F_{H1} which ends at the centre of this circle. F_{H1} is calculated from the positions and weights of the heavy atoms of the first derivative.
3. With the beginning of this vector as origin draw a circle of radius $|F_{PH1}|$. The intersections of the two circles represent the two possible ways for F_P and F_{H1} to add together to give F_{PH1}.
4. Repeat the steps 2 and 3 for other derivatives represented by F_{PH2}, F_{PH3}, and so on.

A diagram of this kind is called a *Harker diagram* after its inventor, David Harker.

Ideally all the circles will intersect at one point which identifies the phase α_P, but in a real Harker diagram, the circles do not intersect exactly, and no phase angle satisfies the isomorphous replacement relationship perfectly. In such a case, the Harker diagram gives a good visual representation of the range of probable phase angles, but does not lend itself to an exact estimate of the phase angle (Fig. 7.21).

Example: Harker diagrams for haemoglobin (Cullis *et al.* 1961)

In presenting the first three-dimensional protein structure determination (at 5.5 Å resolution), six derivatives were used. The authors showed extreme examples of Harker diagrams for some of their phase determinations (Fig. 7.22).

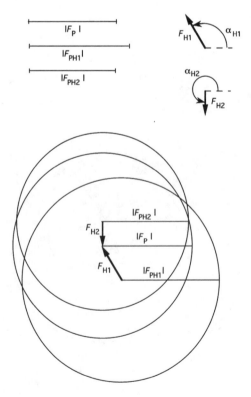

Fig. 7.21 Harker construction for a more typical case.

Computational procedures have been established to calculate a probability distribution for the phase of each reflection (Box 7.8, Fig. 7.25). Reflections of uncertain phase determination will introduce errors into an electron-density map, so they must be weighted down. The appropriate weight, calculated from the phase probability distribution, is called the *figure of merit*. It would be 1 for a perfectly defined phase angle, tending to 0 as the errors increase. The haemoglobin workers published a distribution curve for the figures of merit in their 5.5 Å study, showing a mean figure of merit of 0.78 (Fig. 7.26). Although many phases were accurately determined, there is a long 'tail' of poorly measured phase angles.

For subsequent calculation of electron-density maps, errors are minimized by choice of a 'best' phase from the phase probability distributions, and by weighting down poorly phased reflections by the figure of merit, m. The best estimate for the structure factor F_P is to give it a magnitude $m|F_{obs}|$ and phase α_{best} (Box 7.9).

Performance measures for isomorphous replacement

The mean figure of merit m (taken over all reflections, or of a group of reflections) may be quoted as evidence of the accuracy of phase determination using all the derivatives. Values may be calculated for different resolution ranges, showing how the phase determination deteriorates at higher resolution. These figures are occasionally misleading. If an investigator underestimates the magnitude of his experimental errors, including the errors in calculating F_H, the mean figure of merit will be unrealistically high. Modern computational procedures usually correct this mistake.

It is also important to know how effective a particular isomorphous derivative is, in finding the phase angles. This is done by finding out how accurately one derivative satisfies the isomorphous replacement relationship presented in Box 7.2 and illustrated in Fig. 7.6. The *phasing power* compares the magnitudes of the calculated heavy atom contributions $|F_H|$ to the error in closing the triangle, assuming the best phase, $x(\alpha_{best})$. The mean value of this ratio may also be presented as a function of resolution.

Even after careful reinterpretation of difference maps, the errors in isomorphous replacement phases may be quite large. A particular isomorphous derivative is a very useful aid to

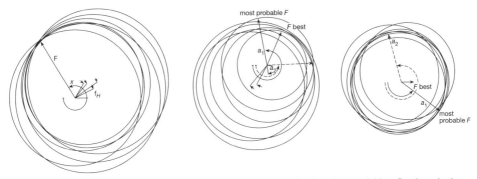

Fig. 7.22 Harker construction for three haemoglobin reflections. In the first, all six derivatives indicate the same phase. The second and third present more typical cases which give only weak phase indications. (Reproduced from Cullis *et al.* (1961), by permission of the Royal Society.)

Box 7.8 Calculation of probable phase angles by the isomorphous replacement method

The Harker diagram shown in Fig. 7.21 is more typical than Fig. 7.20. Because of errors of many kinds, no phase will fit all the observations perfectly. If the true phase were α_P, it would imply an error that can be estimated from the 'lack of closure', x, which is the difference between $|F_{PH}|_{\text{observed}}$ and the value $|F_{PH}|_{\text{calc}}$ calculated from the isomorphous replacement equation (Fig 7.23). Thus

$$|F_{PH}|_{\text{calc}} = |F_P + F_H| = |(|F_P|_{\text{observed}} \exp[i\alpha_P] + F_{H\text{calc}})|,$$

and the lack of closure

$$x(\alpha_P) = |F_{PH}|_{\text{observed}} - |F_{PH}|_{\text{calc}}$$
$$= |F_{PH}|_{\text{observed}} - |(|F_P|_{\text{observed}} \exp[i\alpha_P] + (F_H)_{\text{calc}})|$$

In this equation there are two observed quantities, and the quantity $(F_H)_{\text{calc}}$ is derived from the positions and weights assigned to the heavy atom derivative. The symbol $x(\alpha_P)$ reminds us that this particular value of x is obtained by assuming a phase α_P for $|F_P|$.

The computer works through all the values of α_P from 0 to 2π and calculates a value of x for each. If the lack of closure is small, the phase is more likely to be correct; if it is large, the phase is unlikely. Figure 7.24 shows the results obtained using the data used for the first derivative in Fig. 7.20 as a full line.

Using this information, a 'phase probability curve' may be calculated. This is done by assuming that the probability of a lack of closure error follows a Gaussian distribution, the standard error of which is the quantity E_{iso} mentioned in the main text below. For a single isomorphous derivative the phase probability curve is symmetrical, suggesting two equally possible phase angles (fine full line, Fig. 7.25).

Now, a second isomorphous derivative may be analysed in the same way, generating a different phase probability curve, which is symmetrical about a different phase. The results for the second derivative in Fig. 7.20 are shown as dashed lines in Fig. 7.24 and 7.25. A joint phase probability curve may then be generated by multiplying together the two phase probability curves (thick line, Fig. 7.25). This curve has lost the symmetry that represents the ambiguity of phase given by a single derivative. The results for further derivatives may be incorporated in a similar way, generating a phase probability curve based on all the data for this reflection.

In this idealized example, both derivatives agree on the same phase angle. A similar calculation could be carried out on the more typical Harker diagram of Fig. 7.21, and this will be taken up in Box 7.9.

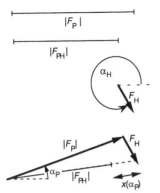

Fig. 7.23 Given $|F_P|$, $|F_{PH}|$, and F_H, every possible phase angle, α_P, implies a corresponding lack of closure $x(\alpha_P)$.

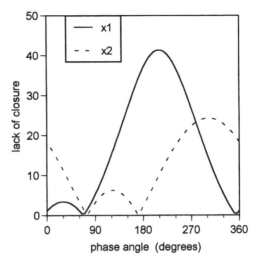

Fig. 7.24 Lack of closure x from the two heavy atom derivatives of the Harker diagrams in Fig. 7.20. In this idealized case, both derivatives give exact closure at a phase angle $\alpha = 70°$, but each gives a second exact closure at some other phase angle.

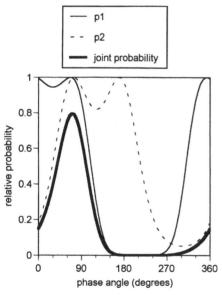

Fig. 7.25 The lack of closure of Fig. 7.24 is used to assign relative probabilities $p1$ and $p2$ from the two isomorphous derivatives. Each relative probability curve is symmetrical and has two peaks representing two equally possible phase angles. Taking both into account, a joint probability curve is generated by multiplying the two probabilities together. (For increased clarity, the joint probability curve has been multiplied by 0.8.) This joint probability distribution shows the preferred phase angle of 70°, but shows phases from 0° to 120° to be quite likely.

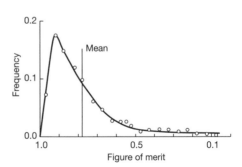

Fig. 7.26 Distribution of figure of merit for haemoglobin phases to 5.5 Å (reproduced from Cullis *et al.* (1962), by permission of the Royal Society).

phase determination if the phasing power exceeds 1.5. It is pretty useless if the phasing power is less than 1.0 (meaning that the heavy atom structure factors are smaller than the inherent errors).

Isomorphous replacement in practice

Phase determination by isomorphous replacement is always severely affected by experimental error, especially the error in assuming the crystals to be truly isomorphous. The effects of lack of isomorphism become more severe at higher resolution. At the same time, disorder in the crystals makes the intensities weaker and less accurately determined at high resolution, and disorder in the heavy atoms makes their effects weaker.

The size of these effects may be monitored by the changes of phasing power and mean figure of merit as a function of resolution. An electron-density map is only as good as the

---

Box 7.9 How to decide the 'best' structure factor to be used for calculation of an electron-density map

The phase probability curve of Fig. 7.25 may be plotted on a circle of radius $|F_P|_{observed}$ and then represents a probability distribution for the structure factor F_P drawn out on an Argand diagram (Fig. 7.27).

To calculate an electron-density map, we have to decide a particular value as F_P for each Bragg reflection. We only have a probability distribution for it, and the best strategy to minimize error in the calculated electron density is to choose the centre of gravity of the distribution. This is shown as F_{best} in Fig. 7.27.

The phase $\alpha_{P,best}$ is known as the 'best phase'. It is evident that the centre of gravity of the phase distribution will always lie inside the circle of radius $|F_P|_{observed}$ so that the magnitude $|F_P|_{best}$ will always be less. This amounts to a weighting factor which reduces the magnitude assigned to this structure factor depending how accurately its phase is known. This magnitude is known as the *figure of merit*, m, where

$$m = \frac{|F_P|_{best}}{|F_P|_{observed}}.$$

It may be shown that if the phase probability curve has only one peak, m represents the cosine of the estimated phase error.

The most important application of this weighting scheme, however, is in cases where the phase probability curve has two peaks, when the 'best' phase may be very different from either of the individual indicated phases. Figure 7.27 was generated from the idealized case of Fig. 7.20, where two isomorphous derivatives agreed exactly upon the phase angle. In practical cases, the agreement between indications from different derivatives is less than perfect, as in Fig. 7.21. Such cases show more clearly the importance of a probability analysis. Figures 7.28–7.30 present results from this example, which gives a phase probability with two peaks. The uncertainty in the phase leads to a much smaller figure of merit, indicated by the small magnitude of $|F_P|_{best}$.

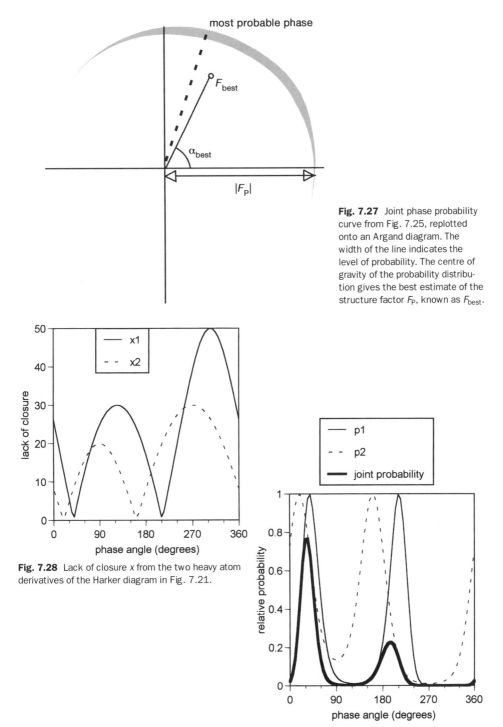

Fig. 7.27 Joint phase probability curve from Fig. 7.25, replotted onto an Argand diagram. The width of the line indicates the level of probability. The centre of gravity of the probability distribution gives the best estimate of the structure factor F_P, known as F_{best}.

Fig. 7.28 Lack of closure x from the two heavy atom derivatives of the Harker diagram in Fig. 7.21.

Fig. 7.29 Relative probabilities for each isomorphous derivative, and joint probability corresponding to Fig. 7.28.

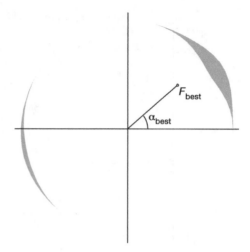

Fig. 7.30 Joint phase probability curve from Fig. 7.29, replotted on to an Argand diagram.

phases that are used to calculate it, and loss of phase accuracy may severely worsen the apparent resolution of the density. Even so, surprisingly good electron-density maps may be obtained from sets of phase angles which have large errors. As a rough guide, electron-density maps become severely inaccurate if m averages less than 0.5, implying a mean phase error exceeding 60°.

Example: good isomorphous derivatives for *Erythrina* trypsin inhibitor (Onesti *et al.* 1991)

For *Erythrina* trypsin inhibitor, the first example given in this chapter, two Pb sites were identified from a difference Patterson function, and difference Fourier maps showed 3, 2, and 2 heavy atom sites, respectively, in the other derivatives. A residual map identified two more Pb sites (four in all). These were used to calculate phase angles by isomorphous replacement with data from all four derivatives to 3.3 Å and using two derivatives to 2.7 Å. Figure 7.31 shows the phasing power of different derivatives and the final mean figures of merit as a function of resolution. The two mercury derivatives give little phase information beyond about 5 Å. The lead derivative is powerful, and appears to improve beyond 3.5 Å. It increasingly dominates over the iridium derivative in phase determination and, by 3.0 Å, there is effectively a single isomorphous derivative, with only weak resolution of the phase ambiguity. For this reason the mean figure of merit decreases below 0.6 at higher resolution. The electron-density map had a mean phase error (compared to the final refined structure) of 57°, but six polypeptide fragments, representing 80% of the molecule, could be assigned confidently. Interpretation was improved using methods described in Chapter 10.

D-glycerate dehydrogenase: poor isomorphous derivatives not enough (Goldberg 1993)

Figure 7.32 shows the phasing power of each of the seven isomorphous derivatives used in the study of D-glycerate dehydrogenase, as a function of resolution. It also shows the

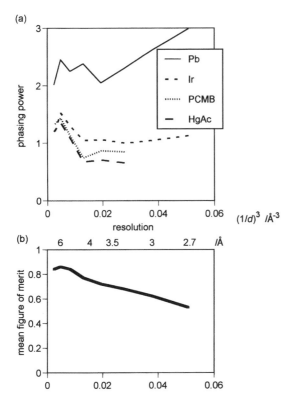

Fig. 7.31 Isomorphous replacement for the *Erythrina* trypsin inhibitor. The horizontal axis includes equal numbers of reflections in equal widths. (a) Phasing power for each derivative. (b) Mean figure of merit using all derivatives. (Reproduced by permission from the *Journal of Molecular Biology*, Onesti *et al.* 1991.)

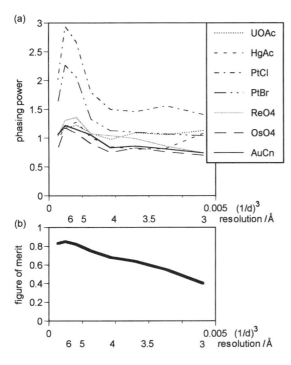

Fig. 7.32 Phasing power for seven derivatives of D-glycerate dehydrogenase as a function of resolution. The horizontal axis includes equal numbers of reflections in equal widths. (a) Phasing power for each derivative. (b) Mean figure of merit using all derivatives. (Data from Goldberg 1993.)

overall figure of merit obtained from these seven derivatives. It is seen that two derivatives provide good phasing power up to about 4 Å resolution, and one of these continues to be reasonably powerful to 3 Å. Two other derivatives have useful phasing power only to 5 Å. The figure of merit is low beyond 3.3 Å.

A further problem in this example is that the two derivatives with best phasing power ($PtCl_4^{2-}$ and $PtBr_4^{2-}$) bind to the protein at precisely the same sites, and do not therefore provide independent information. The phase information from the other derivatives was poor beyond 4 Å, and the 3 Å map generated by multiple isomorphous replacement was not interpretable. A further procedure was needed before interpretable electron density could be obtained.

Conclusion

The aim of this part of a structural investigation is usually to obtain an electron-density map capable of interpretation in terms of atomic positions (Chapter 11), aligned to the amino-acid sequence which is usually known from other techniques. Once this is achieved, methods are available to refine the electron density and improve the resolution (Chapter 12). Poor phase angles may be improved by other methods, discussed in Chapters 9 and 10.

In Chapter 8 a different method is discussed, also based on introducing heavier atoms into a structure, which is often even more powerful than isomorphous replacement.

Further reading

A fairly extensive treatment of the material in this chapter will be found in:

Drenth, J. (1994). *Principles of protein X-ray crystallography*, pp. 129–82. Springer, New York.

Zanotti, G. (1992). Protein crystallography. In *Fundamentals of crystallography* (ed. C. Giacovazzo), pp. 535–97. International Union of Crystallography/Oxford University Press, Oxford.

Recent developments have been reviewed by:

Ke, H. (1997). Overview of isomorphous replacement phasing. *Methods Enzymol.*, **267**, 448–60.

8

Anomalous scattering

There is another technique for phase determination which also depends on small changes to the diffraction caused by unusual (mostly heavy) atoms. This method is now becoming more frequently used than isomorphous replacement.

To introduce it, recall Friedel's law, presented in Chapter 4, which states that F_{hkl} and $F_{-h,-k,-l}$ have the same magnitude, but phases of opposite sign. We are now going to consider cases where Friedel's law is breached. The insight of Johannes Bijvoet (pronounced **By**-foot) was that the breach of Friedel's law could provide phase information. In 1949 he wrote:

There is, in principle, a general way of determining the sign of [a phase angle]...We can use the abnormal scattering of an atom for a wavelength just beyond its...absorption limit....At present we are testing the applicability of this method in an actual analysis. By the above method it also becomes possible to attribute the d or l structure to an optically active compound on actual grounds and not merely by a basic convention.

(Bijvoet 1949)

X-rays set the electronic charges around atoms vibrating, and these vibrations generate radiation of the same frequency, which is propagated in all directions. This is the coherent scattering that gives rise to diffraction effects. Normally, the electrons vibrate in step with the incident beam. If, however, the incident photons have an energy close to a transition energy which can bring the atom to an excited state, the electronic vibration gets out of step. Instead of re-radiating in phase with the incident beam, the radiated energy has a different phase. Also, the intensity of coherent scattering is reduced, because some energy is absorbed to bring about the transition. This effect is called *anomalous scattering*. More detail is given in Box 8.1. Figure 8.1 shows how X-ray absorption varies with wavelength near the transition energy. X-rays become rapidly less penetrating as wavelength increases, but this trend is interrupted by a sharp 'edge' at a wavelength that corresponds to an electronic transition.

In practice, at the wavelengths of X-rays convenient for diffraction experiments (less than 1.6 Å) atoms lighter than phosphorus or sulphur behave as normal scatterers because they have no transitions of corresponding energy. At wavelengths very near the energy of an electronic transition, anomalous scattering can become a significant fraction

Box 8.1 X-ray scattering near an atomic transition energy

The transition energy is the energy needed to throw an electron out of one of the orbitals of an atom, either to eject it totally from the atom, or to move it into an unoccupied orbital of higher energy. Over the wavelengths where a photon has the appropriate energy (see Box 1.1), there is a rapid change in the atomic absorption factor, creating a so-called 'absorption edge' (Fig. 8.1), which spans the range of energies needed to promote an electron from one particular orbital shell, whether to promote into the next higher shell, or to eject the electron completely. The absorption factor, plotted against wavelength, gives a characteristic stepped profile. (The details of this profile are studied in the extended X-ray absorption fine structure (EXAFS) analysis technique.) The whole atomic absorption spectrum of an element in the X-ray region consists of a series of absorption edges corresponding to different shells of its electrons $K, L, M, \ldots$, with principal quantum numbers $1, 2, 3, \ldots$.

Anomalous scattering occurs close to these absorption edges. Since some of the incident radiation is used to create transitions in the atoms, the total coherent scattering at these wavelengths is reduced, and there is also a change in the phase of the coherently scattered X-rays. This phase change can be expressed by saying that the atomic scattering factor has real and imaginary components.

Thus, the atomic scattering factor at any wavelength may be written:

$$f(\lambda) = f_0 - \delta f'(\lambda) + i f''(\lambda), \tag{1}$$

where f_0 is the conventional atomic scattering factor for wavelengths far from an absorption edge, $\delta f'(\lambda)$ is the amount by which the normal scattering is reduced at wavelength λ, and $f''(\lambda)$ is the amount of 'anomalous scattering', the out-of-phase component of the scattering, at this wavelength. The units of $f, f_0, \delta f'$ and f'' are all expressed in electrons.

For compactness, the real part of the atomic scattering factor $(f_0 - \delta f')$ is often written as f', so that (1) may be written briefly as

$$f = f' + i f''. \tag{2}$$

Fig. 8.1 X-ray absorption spectrum of selenium over a range of wavelengths. The *K* absorption edge at 0.98 Å is well placed for crystallographic study. The wavelength of the selenium *L* edge is far too great for useful crystallographic experiments. The *L* edge has three components at 7.5, 8.4 and 8.64 Å. Data from Brookhaven National Laboratory (physics.nist.gov/PhysRefData/).

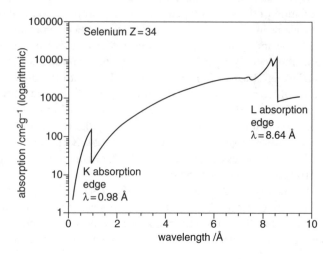

of the total scattering. In this chapter the atoms which are scattering anomalously are referred to as heavy atoms, H.

Useful anomalous scatterers can be far lighter than the heavy elements needed for macromolecular isomorphous replacement. From iron ($Z=26$) to palladium ($Z=46$) the K absorption edges are at convenient wavelengths. Much heavier atoms (the lanthanides and beyond) give strong anomalous effects from the L edges, at useful wavelengths.

Remember that the X-ray scattering power of an atom is measured by comparison with the scattering of an electron. At low scattering angle and 'normal' wavelengths, an atom scatters according to the number of electrons it contains (Box 5.2). At wavelengths close to the absorption edge, the atomic scattering factor includes an 'anomalous' component, shown in Fig. 8.2, and presented in algebraic form in Box 8.2. The anomalous component causes

---

Box 8.2 Anomalous scattering by a heavy atom

Assuming a wavelength where scattering is normal, we wrote

$$F_H = \sum_M f_H \exp[2\pi i \mathbf{h} \cdot \mathbf{x}_{H,i}] \exp[-B_{H,i} \sin^2\theta/\lambda^2],$$

for the structure factor of the heavy atoms (see eqn 2, Box 7.2). The summation is taken over each of the M 'heavy' atoms H.

At a wavelength where anomalous scattering occurs, f_H becomes a complex quantity ($f'_H + if''_H$) (Box 8.1, eqn 2). Usually only one kind of anomalous scatterer is used, so the same value of f_H applies to all the anomalous scatterers. Close to an absorption edge, the 'normal scattering component' f'_H is significantly reduced from its value f_H for normal scattering. The second quantity if''_H, the 'anomalous component', is an *imaginary* quantity, that is to say its phase is $\pi/2$ (90°). Though it is small at wavelengths longer than the absorption edge, it may have a large positive value at shorter wavelengths close to the edge. In symbols:

$$F_H(\lambda) = \sum (f'_H + if''_H) \exp[2\pi i \mathbf{h} \cdot \mathbf{x}_{H,i}] \exp[-B_{H,i} \sin^2\theta/\lambda^2].$$

The symbol (λ) is included to remind us that the actual values will depend upon the wavelength. If we know the positions $\mathbf{x}_{H,i}$ of the heavy atoms, and have good data on the values of f'_H and f''_H at the appropriate wavelength, a good estimate of F_H can be made.

We will call the normal scattering by the heavy atom F_{HN} (Fig. 8.2). This normal scattering is produced by the normal components of the atomic scattering factor, f'_H:

$$F_{HN}(\lambda) = \sum f'_H \exp[2\pi i \mathbf{h} \cdot \mathbf{x}_{H,i}] \exp[-B_{H,i} \sin^2\theta/\lambda^2].$$

The anomalous component, F_{HA} depends upon the anomalous component of the atomic scattering factor:

$$F_{HA}(\lambda) = \sum if''_H \exp[2\pi i \mathbf{h} \cdot \mathbf{x}_{H,i}] \exp[-B_{H,i} \sin^2\theta/\lambda^2]. \tag{1}$$

These summations are taken over all the anomalously scattering atoms H. The total scattering by these atoms is:

$$F_H(\lambda) = F_{HN}(\lambda) + F_{HA}(\lambda).$$

These relationships are demonstrated in Fig. 8.3.

scattering which is out of phase with the incident X-ray beam. At these wavelengths, the 'normal' component of the scattering is reduced.

Violation of Friedel's law

Friedel's law states that a pair of reflections h,k,l and $-h,-k,-l$ have the same structure amplitude, and that the phases have opposite sign (say $+\alpha$ and $-\alpha$). But at a wavelength close to the absorption edge of an atom, its scattering does not obey this rule. Comparing Friedel-related reflections, the heavy atom structure factors $F_H(h,k,l)$ and $F_H(-h,-k,-l)$ have the same magnitudes, but they do not have opposite phases (Fig. 8.3). Although the normal parts of the scattering factors have opposite phases, the anomalous parts behave in a different way, shown in Figs 8.4 and 8.5.

Friedel's law is based on the assumption that atoms scatter normally. It states that the structure factor $F(h,k,l)$ has the same magnitude as $F(-h,-k,-l)$, but opposite phase. We have seen that atoms which scatter anomalously do not obey this phase relationship. Bijvoet realized that when a crystal contains both normal scatterers as well as one or more heavy atoms which are scattering anomalously, there would be an observable effect. As explained in the following example, the violation of Friedel's law can be observed directly from diffraction measurements, because the scattered intensities $(F_H(h,k,l))^2$ and $(F_H(-h,-k,-l))^2$ are different.

normal scattering
(wavelength λ_N)

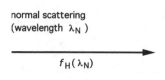

anomalous scattering
(wavelength λ_A)

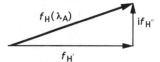

Fig. 8.2 A heavy atom scattering factor f_H at two different wavelengths. At wavelength λ_N it scatters normally. At wavelength λ_A there is an anomalous scattering component f''_H, and also the normal component of the scattering is reduced to f'_H.

Example: absolute configuration of the tartrate ion (Bijvoet *et al.* 1951)

Bijvoet exploited this effect to determine the absolute configuration of the molecules in crystals of sodium rubidium D-tartrate. Sodium and the atoms of the tartrate ion behaved as normal scatterers in his experiment, but the rubidium ion was a significant anomalous scatterer. When the anomalous scattering by the rubidium ions was added to the normal scattering, the anomalous rubidium scattering caused the magnitudes of the total structure factors

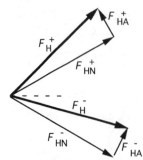

Fig. 8.3 Normal and anomalous components of structure factors in a Friedel pair.

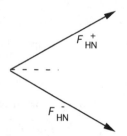

Fig. 8.4 The normal components are related as complex conjugates, by reflection across a horizontal line.

Fig. 8.5 The anomalous components are related by reflection across a vertical line. The negative of one is the complex conjugate of the other.

to be different, for Friedel-related pairs of reflections (Fig. 8.6). In some cases the intensity $I(h,k,l)$ exceeded $I(-h,-k,-l)$ and sometimes it was less. Bijvoet used these observations to determine which of the two possible tartrate enantiomers (representing D-tartrate and L-tartrate) was consistent with the observed intensity differences.

The 'Fischer convention' had been adopted arbitrarily by Emil Fischer to present the conformations of organic molecules, particularly sugars, in a consistent way. Bijvoet's structure of the tartrate ion demonstrated that Fischer had actually made the correct choice for the 'handedness' at asymmetric carbon atoms. The difference of scattered intensity between I_{hkl} and $I_{-h,-k,-l}$ is known as the Bijvoet difference, and the difference between $|F_{hkl}|$ and $|F_{-h,-k,-l}|$ will be called the Bijvoet amplitude difference.

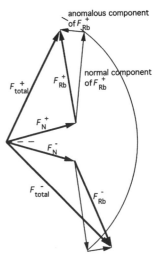

Fig. 8.6 The total scattering by sodium rubidium tartrate crystals is the sum of the scattering by 'normally scattering atoms' and the scattering by rubidium which has an anomalous component. This leads to a difference in the magnitude of the total scattering for the h,k,l reflection identified by +, and that for $-h,-k,-l$ identified by −. As emphasized by the arc, $|F_{total}^{+}|$ differs from $|F_{total}^{-}|$.

Anomalous scattering in macromolecular crystals

Tunable synchrotron wavelengths make the anomalous scattering method much more powerful, because measurements can be made at a precisely chosen wavelength. Figure 8.7 shows the variation of the normal and anomalous components of scattering as the wavelength is scanned through the energy of an atomic transition. The anomalous component of the atomic scattering factor f'' closely follows the variation of the X-ray absorption coefficient.

Anomalous scattering effects are smaller than those obtained by isomorphous replacement, but can be measured more accurately because measurements are made on the same crystal, and interpreted more accurately, because there is no question of 'non-isomorphism'.

The anomalous scattering method of phase determination is based on the following idea:

The total scattering by the structure is the sum of the normal scattering and the anomalous scattering. But anomalous scattering has opposite effects on a reflection and on its Friedel-related partner. Phase information may be obtained from measurement of $|F_{PH}|$ for each member of the Friedel pair.

Details are given in Box 8.3, and Fig. 8.8 shows the structure factor of the complex of a normally scattering protein, P, with an anomalous scatterer, H. The figure shows the structure factors F_{PH}^{+} for reflection h,k,l and F_{PH}^{-} for $-h,-k,-l$. To make the comparison more graphic, Fig. 8.9 shows the scattering for $-h,-k,-l$ reflected through the horizontal 'real' axis. All the normal components of scattering become identical, while the anomalous component F_{HA}^{-} for $-h,-k,-l$ is reversed. The anomalous component fulfils a role similar to F_H in isomorphous replacement. The difference between F_{PH}^{+} and F_{PH}^{-} is $2F_{HA}$. Figure 8.10 shows how this method gives, in practice, two possible phase angles for F_P, just as a single isomorphous derivative does.

Box 8.3 An anomalous scattering equation for phase determination

Scattering by the light atoms of the structure is usually assumed to be normal, and most often this includes sulphur atoms (although this may be a significant approximation). If 'heavy' atoms are present which are significant anomalous scatterers, their scattering will include a normal component F_{HN} and an anomalous component F_{HA} (Box 8.2).

The total normal scattering, F_N, includes all the scattering by light atoms, and also the normal component of the scattering by the anomalous scatterers. The total scattering for reflection $\boldsymbol{h}$ is

$$F(\boldsymbol{h}) = F_N(\boldsymbol{h}) + F_{HA}(\boldsymbol{h}). \tag{1}$$

(Only the H atoms are anomalous scatterers, so the anomalous component is written F_{HA}.) Now consider scattering for the reflection $-\boldsymbol{h}$. The normal scattering obeys Friedel's law, and

$$F_N(-\boldsymbol{h}) = F_N(\boldsymbol{h})^\star. \tag{2}$$

Figure 8.5 shows graphically that the corresponding relationship for the anomalous parts of the structure factors, F_{HA}, is precisely the negative of Friedel's law,

$$F_{HA}(-\boldsymbol{h}) = -F_{HA}(\boldsymbol{h})^\star. \tag{3}$$

(This result is derived algebraically in Box 8.4.) Writing eqn 1 for reflection $-\boldsymbol{h}$ gives

$$F(-\boldsymbol{h}) = F_N(-\boldsymbol{h}) + F_{HA}(-\boldsymbol{h}), \tag{4}$$

and using eqns 2 and 3 this becomes

$$F(-\boldsymbol{h}) = F_N(\boldsymbol{h})^\star - F_{HA}(\boldsymbol{h})^\star,$$

or its complex conjugate

$$F(-\boldsymbol{h})^\star = F_N(\boldsymbol{h}) - F_{HA}(\boldsymbol{h}). \tag{5}$$

Finally, by subtracting eqn 5 from eqn 1, one gets

$$F(\boldsymbol{h}) - F(-\boldsymbol{h})^\star = 2F_{HA}(\boldsymbol{h}), \tag{6}$$

as shown graphically in Fig. 8.9. The important point to bear in mind now is that $|F(\boldsymbol{h})|$ and $|F(-\boldsymbol{h})|$ are measurable quantities, and if the positions of the anomalous scatterers, H, are known, F_{HA} can be calculated from Box 8.2, eqn 1. (To calculate F_{HA}, the wavelength must be known accurately and used to obtain f' and f'' from tabulated values.) Eqn 6 is an anomalous scattering equation, which can be used just like the isomorphous replacement equation (Box 7.2, eqn 3). The magnitudes of two of its components are measurable, while the third quantity may be calculated.

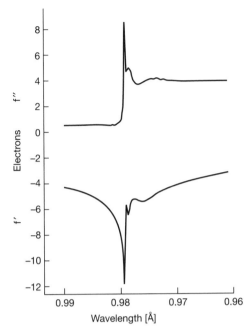

Fig. 8.7 Variation of the normal scattering and the anomalous scattering of selenium as a function of wavelength, going through the *K* absorption edge (reproduced from Hendrickson *et al.* (1990), by permission of the European Molecular Biology Organisation). The quantity plotted as f' is the loss of normal scattering power, $f'_{Se} - f_{Se}$ or $-\delta f'$ in the notation used here.

Box 8.4 The anomalous component of $F(-\boldsymbol{h})$

The anomalous component of scattering for reflection $-\boldsymbol{h}$ may be calculated from eqn 1, Box 8.2. First observe that multiplying this equation by $-i$ gives:

$$-iF_{HA}(\boldsymbol{h}) = \sum_{H} f''_H \exp[2\pi i \boldsymbol{h} \cdot \boldsymbol{x}_{H,i}] \exp[-B_{H,i} \sin^2\theta/\lambda^2]. \tag{1}$$

For reflection $-\boldsymbol{h}$, the corresponding expression is:

$$-iF_{HA}(-\boldsymbol{h}) = \sum_{H} f''_H \exp[-2\pi i \boldsymbol{h} \cdot \boldsymbol{x}_{H,i}] \exp[-B_{H,i} \sin^2\theta/\lambda^2]$$

$$= \sum_{H} f''_H \exp[2\pi i \boldsymbol{h} \cdot \boldsymbol{x}_{H,i}]^\star \exp[-B_{H,i}\sin^2\theta/\lambda^2]. \tag{2}$$

Since the right-hand sides of eqns 1 and 2 are complex conjugates, the same must be true of the left-hand sides, so

$$-iF_{HA}(\boldsymbol{h}) = (-iF_{HA}(-\boldsymbol{h}))^\star = i(F_{HA}(-\boldsymbol{h}))^\star.$$

Multiplying this expression by i gives:

$$F_{HA}(\boldsymbol{h}) = i(-iF_{HA}(-\boldsymbol{h}))^\star = -(F_{HA}(-\boldsymbol{h}))^\star. \tag{3}$$

Compared to Friedel's law in Box 8.3, eqn 2, this relationship is exactly the negative. The result is illustrated graphically in Fig. 8.5.

Combining anomalous scattering observations with isomorphous replacement

If the anomalously scattering atoms are the same as the heavy atoms, H, used in an isomorphous replacement, the anomalous component F_{HA} differs in phase by 90°, compared to the normal scattering F_{HN} of the heavy atoms. Using isomorphous replacement leads to an ambiguity in phase (Fig. 8.11), but this ambiguity is perpendicular to the ambiguity produced from anomalous scattering (Fig. 8.10). Thus, a single isomorphous derivative, where the heavy atom is an anomalous scatterer, and where the Bijvoet difference is measured, can give unique phase determination (Fig. 8.12). This method of phase determination is 'single isomorphous replacement with anomalous scattering' or SIRAS.

An even more powerful method is 'multiple isomorphous replacement with anomalous scattering' (MIRAS), in which several different heavy atoms are substituted isomorphously into a protein crystal and Bijvoet differences are measured for each. It has become a routine matter to make separate intensity observations of the members of a Bijvoet pair, and to use the additional observations to improve the accuracy of phase determination.

Example: galactose oxidase (Ito *et al.* 1991)

Carefully documented data are available (Ito 1991) for the structure determination of galactose oxidase (molecular weight 68 000) using a parent and three heavy atom derivatives, using both isomorphous replacement methods and anomalous scattering methods for phase determination (Ito *et al.* 1991).

The three derivatives were first evaluated as three separate SIRAS determinations. A mean figure of merit was obtained for each derivative, and the phases were compared with those calculated from the final structure. Table 8.1 shows that the Ir derivative was

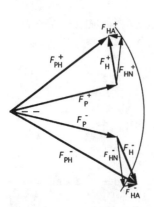

Fig. 8.8 Structure factors for a protein crystal containing anomalously scattering heavy atoms, for a Friedel-related pair of reflections + and −. The diagram is very similar to the earlier figure for sodium rubidium tartrate, but since the heavy atom scattering is usually much weaker than that of a protein, the proportions are altered.

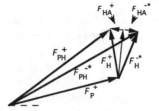

Fig. 8.9 The structure factor $F_{\overline{PH}}$ from the previous figure is plotted here as its complex conjugate $(F_{\overline{PH}})^*$. This makes its normal components exactly equal to the normal components of F_{PH}^+, but the anomalous component $(F_{\overline{HA}})^*$ is exactly the opposite of F_{HA}^+.

substantially better than the others. The phasing power and the mean phase error agree well on the relative powers of the three derivatives.[1] This is also shown clearly by the phasing powers of the three derivatives, plotted against resolution in Fig. 8.13. The phase determination is deteriorating at the resolution limit of 2.5 Å, but the Ir derivative remains powerful.

To show how MIRAS improves on the best available SIRAS results, phases were determined using MIRAS for different combinations of derivatives (Fig. 8.14). MIRAS gives a substantially increased figure of merit over those from the various SIRAS phases, at all resolutions (Fig. 8.15).

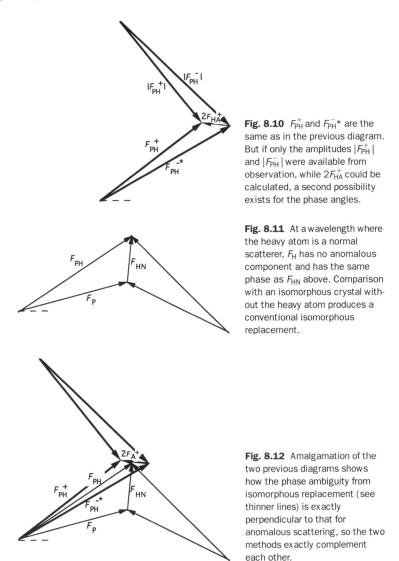

Fig. 8.10 F_{PH}^+ and $F_{\overline{PH}}^*$ are the same as in the previous diagram. But if only the amplitudes $|F_{PH}^+|$ and $|F_{\overline{PH}}|$ were available from observation, while $2F_{HA}^+$ could be calculated, a second possibility exists for the phase angles.

Fig. 8.11 At a wavelength where the heavy atom is a normal scatterer, F_H has no anomalous component and has the same phase as F_{HN} above. Comparison with an isomorphous crystal without the heavy atom produces a conventional isomorphous replacement.

Fig. 8.12 Amalgamation of the two previous diagrams shows how the phase ambiguity from isomorphous replacement (see thinner lines) is exactly perpendicular to that for anomalous scattering, so the two methods exactly complement each other.

[1] The relation between mean phase error and mean figure of merit mentioned in Box 7.9 does not apply in this study for reasons discussed by Ito (1991).

Table 8.1 Statistics on SIRAS phasing for three derivatives of galactose oxidase (data from Ito 1991)

Derivative	K_2PtCl_4	H_2IrCl_6	$Pb(NO_3)_2$
R_{deriv}	0.161	0.149	0.139
Mean figure of merit	0.32	0.41	0.30
Mean phase error	64.6°	52.6°	63.1°

Includes all reflections 20 Å to 2.5 Å.

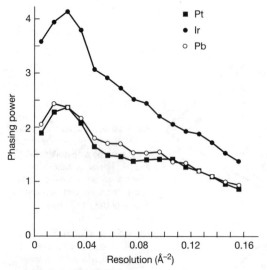

Fig. 8.13 Phasing powers of the three individual isomorphous derivatives, using SIRAS, as a function of resolution (reproduced by permission from Ito 1991).

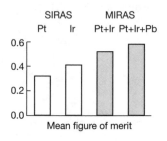

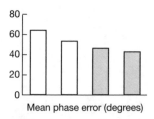

Fig. 8.14 Comparison of mean figure of merit and mean phase error using SIRAS or MIRAS for various combinations of derivatives of galactose oxidase (data from Ito *et al.* 1991).

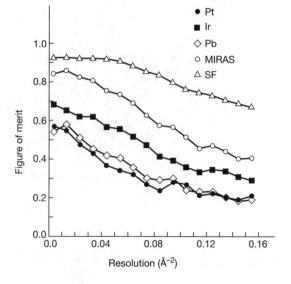

Fig. 8.15 Variation of mean figure of merit with resolution, for each of the three derivatives using SIRAS, and using MIRAS with the pooled data. The data indicated by triangles, marked SF, show further improvement achieved by the solvent-flattening procedure, to be described in Chapter 10. (Reproduced by permission from Ito 1991.)

Phase determination by anomalous scattering only

An increasingly popular method exploits the changes in both the normal scattering (f'_H) and the anomalous scattering (f''_H) of the anomalous scatterers at several wavelengths close to an absorption band. It can be used on its own, without an isomorphous crystal lacking the anomalous scatterer. This method is called 'multiple wavelength anomalous dispersion' (MAD). The great advantage of this method is that only one crystal is needed and there are no errors due to 'non-isomorphism'.

At each wavelength the F_{HN} and F_{HA} components are different, leading to different intensities. It is usual to observe at three different wavelengths, known as 'peak', 'edge', and 'remote' wavelengths, which give significantly different results. At the 'peak' wavelength, f'' should have its largest possible value; at the 'edge' wavelength, the normal part of the atomic scattering factor f' is reduced to its lowest possible value. At the 'remote' wavelength, significantly shorter than these, f' is close to its 'normal' value, although the anomalous component f'' may be quite significant (Fig. 8.16). The two longer wavelengths give data rather similar to two isomorphous derivatives, while the 'remote' wavelength is like a parent. Sometimes a second 'remote' wavelength, a longer wavelength, is also used.

Because anomalous effects are relatively small, accurate intensity measurements are important, indicated by low values of R_{sym} or R_{merge}. But because there are no problems of non-isomorphism (the same crystal is usually used for measurements at all the different wavelengths), the method can provide powerful phase determination. The magnitude of the anomalous differences may be indicated by an R factor, R_{anom}, expressing the magnitude of Bijvoet amplitude differences,

$$R_{anom} = \frac{\text{mean Bijvoet amplitude difference}}{\text{mean amplitude}}.$$

Useful phase indications can be obtained if $2R_{anom}$ is significantly greater than R_{merge}. (The factor 2 derives from the fact that R_{merge} is based on comparison of intensities, see Box 6.1.)

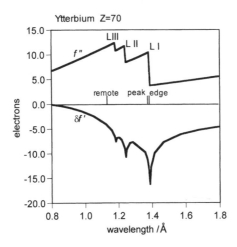

Fig. 8.16 Anomalous scattering factor f'' and reduction of normal scattering factor $\delta f'$ for ytterbium, showing the three L absorption edges. Wavelengths used in the example are indicated. Atomic scattering factor data from Brookhaven National Laboratory (physics.nist.gov/PhysRefData/).

Heavy atom positions may be found from anomalous difference Patterson functions in a way analogous to that described for isomorphous replacement (Chapter 7). A Patterson function is calculated, the coefficients of which are the squares of the Bijvoet amplitude differences $(|F^+| - |F^-|)^2$. This function has peaks at the vectors between anomalously scattering atoms.

Example of MAD: MRP8 calcium-binding protein (Ishikawa *et al.* 2000)

MRP8 is a small protein with a molecular weight of 11 000. An interesting comparison is given of the power of MAD, using ytterbium as anomalous scatterer, with SIRAS using xenon. Wavelengths 1.38525 Å (peak), 1.38580 Å (edge), and 1.10000 Å (remote) were used for MAD data collection, and 1.0000 Å was used for the SIRAS measurements, both to 2.3 Å resolution.

Table 8.2 shows the extremely powerful phasing achieved by the MAD technique. The comparison with SIRAS is probably exaggerated because the xenon derivative was poorly substituted (R_{deriv} was only 0.10) and intensity measurement was less accurate for both parent and xenon derivative at the resolution limit (see Table 8.3). Nevertheless, the better performance of MAD is impressive.

The summary of data measurement statistics in Table 8.3 shows that high-quality intensity measurements could be made from the ytterbium derivative, until R_{merge} increased severely at the resolution limit. The R factors R_{anom} and R_λ are also given. R_λ is

$$\frac{\text{mean amplitude difference compared to remote wavelength}}{\text{mean amplitude at the remote wavelength}}.$$

It is the analogue of R_{deriv}, comparing data at different wavelengths rather than comparing different isomorphous derivatives. The anomalous differences in the MAD technique are of similar magnitude to the SIRAS isomorphous differences from the poorly substituted xenon derivative.

Selenomethionine in proteins

An important technique exploits the considerable anomalous scattering effects caused by selenium at appropriate wavelength (Fig. 8.7). Selenomethionine has similar chemical

Table 8.2 Comparison of SIRAS phasing (using xenon) with MAD phasing (using ytterbium) in MRP8 (data from Ishikawa *et al.* 2000)

	Phasing power		Figure of merit
	Isomorphous	Anomalous	
SIRAS (with Xe)	2.27	0.48	0.29
MAD (with Yb)			
peak	19.5	13.8	
edge	15.8	8.1	
remote	—	6.2	
overall			0.88

All figures apply only to measurements on non-centrosymmetric reflections.

Table 8.3 Data measurement statistics for MRP8 (data from Ishikawa *et al.* 2000)

	Parent	Ytterbium			Xenon	
		Peak	Edge	Remote	Crystal 1	Crystal 2
$I/\sigma(I)$						
overall	10.1	9.2	9.8	8.3	6.3	9.9
at resolution limit	2.1	3.2	3	2.7	2.1	2.8
R_{merge}						
overall	0.062	0.069	0.068	0.083	0.105	0.062
at resolution limit	0.334	0.226	0.243	0.257	0.359	0.272
R_{anom}						
overall		0.097	0.085	0.085		
at resolution limit		0.155	0.153	0.137		
R_λ						
overall		0.082	0.113			
R_{iso}						
overall					0.103	0.094

properties to methionine, and has the same shape and volume. Almost without exception, it may be incorporated into a protein instead of methionine without destabilizing it. Thus, gene expression in a methionine auxotroph, where the medium contains only seleno-methionine, can replace all the methionine in the expressed protein by selenomethionine.

Example: ribozyme from hepatitis delta virus

The structure of a ribozyme from hepatitis delta virus containing 85 nucleotides, complexed to a small protein RBD, was determined by MAD (Ferré-D'Amaré *et al.* 1998). The RBD protein was expressed so its four methionine residues became selenomethionine, and X-ray intensities to 2.9 Å resolution at remote, peak, and edge wavelengths appropriate for seleni-um were measured from one crystal. Another crystal gave a further set of intensity data (here called 'distant') at a longer wavelength of 1.14 Å. An anomalous difference Patterson function revealed four selenium sites (one was disordered). Figure 8.17 presents the data quality and the phasing power provided by each of the derivatives. The authors present two contributions to the phasing power. One, called 'ano', is derived from the Bijvoet differences; the other, which they call 'iso', is derived from comparison of different wavelengths which have different F_H' values. For the latter, the short-wavelength remote data was treated as 'parent'. As displayed in Fig. 8.7, anomalous scattering was significant at this short 'remote' wavelength (0.976 Å), which provides powerful 'ano' phasing.

Figure 8.17a shows satisfactory overall values of R_{sym} and mean intensities over 20 standard deviations for the three sets of measurements from the first crystal. The second crystal ('distant' data) was slightly less satisfactory. Figure 8.17b demonstrates that inten-sity measurement became poor in the outer shell, with R_{sym} greater than 0.3 (30%).

At the three MAD wavelengths, phasing power is good overall (Fig. 8.17c). The 'ano' phasing power from the Bijvoet differences is greater than the 'iso' phasing power from comparing different wavelengths. As in other MAD studies, the phasing powers compare

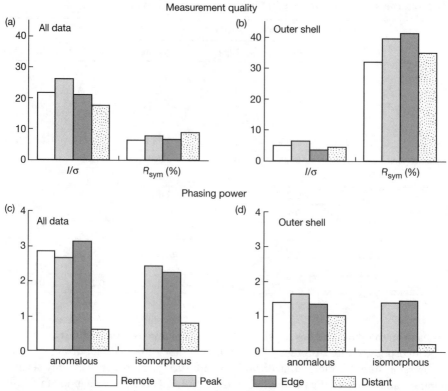

Fig. 8.17 Measurement quality and phasing power at four wavelengths for the hepatitis delta virus ribozyme/RBD complex. Measurement quality (a) all data, (b) outer shell. Phasing power (c) all data, (d) outer shell. The outer shell covers 3.2 to 2.9 Å resolution. (Data from Ferré d'Amaré *et al.* 1998, Table 1.)

favourably with those obtained in many isomorphous replacement studies. This is emphasized by the phasing power at the remote wavelength, where anomalous effects are small. In the outer shell (Fig. 8.17d), the phasing powers from the MAD wavelengths are still fairly useful, but the 'distant' data are contributing very little.

MAD is becoming the most frequently used technique for phase determination by experiment, and the trend to exploitation of anomalous scattering techniques may be expected to increase.

Further reading

Introduction to anomalous scattering methods in proteins:

Fourme, R., Shepard, W., Schiltz, M., Ramin, M., and Kahn, R. (2000). Experimental determination of structure factor phases in biocrystallography. In *Structure and dynamics of biomolecules* (ed. E. Fanchon, E. Geissler, J.-L. Hodeau, J.-R. Regnard, and P.A. Timmins). Oxford University Press, Oxford.

Hendrickson, W.A. (1991). Determination of macromolecular structures from anomalous diffraction of synchrotron radiation. *Science*, **254**, 51–8.

Matthews, B.W. (2001). Anomalous dispersion. Heavy atom location and phase determination with single-wavelength diffraction data. In *International tables for crystallography*, Vol. F. (ed. M.G. Rossmann and E. Arnold), pp. 293–8. International Union of Crystallography/Kluwer, Dordrecht.

Method of phase determination from multiwavelength anomalous data:

Hendrickson, W.A. and Ogate, C.M. (1997). Phase determination from multiwavelength anomalous diffraction measurements. *Methods Enzymol.*, **276**, 494–523.

Applications of selenomethionine:

Doublié, S. (1997). Phase determination from multiwavelength anomalous diffraction measurements. *Methods Enzymol.*, **276**, 523–30.

Hendrickson, W.A., Horton, J.R., and LeMaster, D.M. (1990). Selenomethionyl proteins produced for analysis by multiwavelength anomalous diffraction (MAD): a vehicle for direct determination of three-dimensional structure. *EMBO J.*, **9**, 1665–72.

9

Molecular replacement

The methods described in the Chapters 7 and 8 are the only ones practicable when the structure under study is completely unknown. But as soon as a few structures were known, they suggested that related problems might be tackled in a more direct way. For example, if I know the molecular structure of horse oxyhaemoglobin, surely it can help me in the study of horse deoxyhaemoglobin crystals. Also it should help me to tackle the crystal structure of human oxyhaemoglobin. And, by the way, shouldn't the first protein crystal structure at atomic resolution, that of sperm-whale myoglobin, help me with all of these?

Well-developed techniques are now available to apply these ideas. They will be presented first for the simplest case. Suppose I know accurately the structure of a protein crystal form, A, and I can use this crystal structure to define a molecule M. This requires decisions about the boundaries that identify a molecule in A. Now I have crystallized the protein in a different crystal form, X, the diffraction intensities of which I have measured. How can the crystal structure of X be determined? From what I know of protein structure, it is fair to assume that the two molecular structures are very similar. M can be used as a model for the structure which may be in X.

There is a very simple case where the crystal symmetry is unchanged and the unit cells are virtually the same. Such cases may be studied by electron-density difference maps, similar to those described for the assignment of heavy atom positions in Chapter 7. This is quite common, for example, in the case of single-site amino-acid substitutions in a protein, and the difference map may show the electron-density differences at the substituted amino acid, while the absence of significant density in the rest of the map could confirm that negligible changes had occurred in the rest of the structure.

In the more important case, the crystal forms A and X crystallize differently, and the space-group symmetry is probably different. It is important to begin by determining the symmetry and cell dimensions of the X crystals.

With this information, we can check that the asymmetric unit of X contains only one M-like molecule. The unit cell volume of X, and the molecular weight of the protein that composes it, allow a Matthews volume (Chapter 5) to be calculated, assuming only one molecule in the crystal asymmetric unit, as

$$\frac{\text{volume of unit cell}}{(\text{no of asymmetric units in unit cell}) \times (\text{molecular weight})}.$$

If this comes within the usual range of Matthews volumes, there is likely to be only one M-like molecule in the crystal asymmetric unit of X. (If this is not so, the problem is more complicated, see Chapter 10.)

Solving the problem can then be broken down into four steps:

1. To find the angular orientation of an M-like molecule in the crystal structure of X.
2. Given this angular orientation, to find what point in X corresponds to a chosen origin position for M.
3. To use this information to generate a hypothetical molecular structure of X, composed of M-like molecules.
4. To improve this hypothetical structure, using diffraction measurements from X.

This chapter discusses the first three steps. Step 4 leads into the subject matter of Chapter 12.

Rigid displacements

We are assuming that the structure of X can be represented, well enough, by a structure composed of M-like molecules which are rigidly displaced to the correct positions. Figure 9.1 shows a rigid object displaced from one position to another. The word 'rigid' means that the movement has not distorted it in any way. How can this displacement be described?

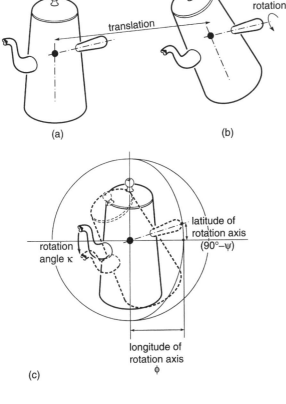

Fig. 9.1 A rigid non-symmetrical three-dimensional object, displaced from one orientation (a) to another (b). Note that the handle of the coffee pot is in the same direction in both positions. It has been rotated about this axis, and translated. (c) The angles that specify the rotation. (Drawn by Neil Powell.)

To deal with the problem quantitatively, we must start by deciding on a reference point (the 'origin'), and defining a system of coordinates. Having done this, it is best to deal with the problem in two steps, called 'rotation' and 'translation':

1. Rotate the original molecule about the reference point until it is exactly parallel to the displaced molecule.
2. Move it, without any rotation, to the displaced position.

To specify the rotation in step 1, three angles must be given. One way is to specify the direction of the axis of rotation by two angles corresponding to the latitude and longitude of the axial direction. A third angle specifies the magnitude of the angle by which it is turned about this axis (Fig. 9.2). By varying these three angles, the molecule can be brought into any desired angular orientation.

Step 2 defines the movement (translation, without rotation) that the molecule has to make in three-dimensional space, after the rotation of step 1, before it arrives at the position of the displaced molecule. Naturally, this movement also requires three values to be specified, representing the displacement of the molecule in each of three directions.

The seemingly simple displacement of a rigid molecule therefore requires six quantities to specify it. To search for the correct displacement, a six-dimensional search is required—a demanding and expensive problem. For example, even if only 10 values were considered for each variable (a very coarse search) this would generate a million (10^6) search points. If a finer search was needed with 100 possible values for each variable, this would generate 10^{12} (a million million, sometimes called a trillion) points to be examined.

Fortunately, the special properties of the Patterson function allow the searches for step 1 and step 2 to be done separately, breaking the six-dimensional search into two three-dimensional ones.

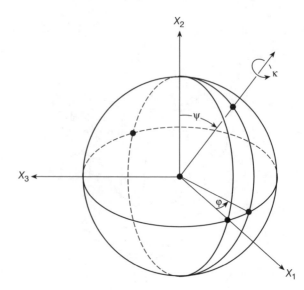

Fig. 9.2 The rotation operation may be specified, for example, by stating a latitude $(90° - \psi)$ and longitude ϕ which define the direction of the rotation axis, and the angle κ through which it is rotated. (Reproduced from Rossmann and Blow (1962) by permission of the International Union of Crystallography.)

Rotational relationships

The Patterson function records the array of interatomic distances in the molecule, described as its self-convolution in Chapter 4. The Patterson function does not depend on any choice of origin in the molecule. The origin of the Patterson function is the point where vectors of zero length are found. It is this feature that allows it to be used to search for a rotation, independently of any translations that may have occurred.

The principle of the *rotation function*, used in a Patterson search, can be understood more simply in two dimensions. In two dimensions, rotation is only possible about an axis perpendicular to the two-dimensional plane, and all possible rotations can be described by stating a single angle between 0° and 360°.

In Fig. 9.3 a simple molecule M, composed of three atoms, is shown. Figure 9.4 shows its Patterson function. If the molecule is rotated (Fig. 9.5) the effect is to rotate its Patterson function through the same angle. Apart from this rotation, the two Patterson functions are identical.

Rotational search in a crystal: simplified case

The real situation is a bit more complicated because X is an unknown crystal structure, not a single molecule. Let us assume that X contains the same molecules as M, but they are rotated as shown in Fig. 9.5. Taking a two-dimensional example, and choosing the simplest possible symmetry, the structure of the crystal X is shown in Fig. 9.6. Here the only symmetry operations are the two two-dimensional lattice translations. For this simple symmetry the Patterson of the crystal X (Fig. 9.7) is like the Patterson function of the isolated molecule, but it is complicated by the existence of lattice translations. It is (see Chapter 4) the convolution of the Patterson of the isolated molecule with the lattice function. It already looks rather complex. Figure 9.8 shows how the Patterson of the isolated molecule can be found in the Patterson function of the crystal.

These figures have shown how the Patterson of the model molecule M can be related to the Patterson function of the unknown crystal structure X, which contains similar molecules. The molecular Patterson has to be rotated through the angle which relates the orientation of molecules in the two structures.

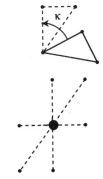

Fig. 9.3 A molecule, M, composed of three atoms.

Fig. 9.4 The Patterson function of the three atoms of M. Three vectors of zero length are at the origin.

Fig. 9.5 A similar molecule rotated through an angle κ, and its Patterson function. It is a rotated version of the Patterson of M.

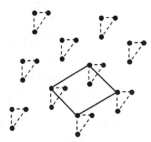

Fig. 9.6 The crystal structure of X, composed of the same molecules arranged in a simple lattice. A unit cell is outlined.

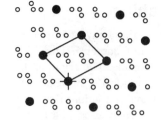

Fig. 9.7 The corresponding Patterson function. The three atoms in the crystal unit cell give a triple-weight vector at the origin of the Patterson function and at the other lattice points. There are also six single-weight vectors per cell.

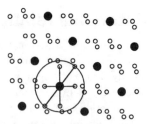

Fig. 9.8 The rotated Patterson function of M, laid on the Patterson of X, picks out six of the Patterson vectors of the crystal. They all lie within a circle, the radius of which is the maximum dimension of M.

When the rotated Patterson function of molecule M is superimposed on the Patterson function of crystal X, it picks out the parts of the Patterson of X which arise from intramolecular vectors. It does not match the rest of the function (Fig. 9.8).

In practice, I do not know how the molecules M are oriented in the crystal structure X. I have to search through all possible orientations of M to find a good fit. As shown in Fig. 9.8, it is the intramolecular vectors that will provide the fit. All the peaks that fit must be within a circle, the radius of which is the maximum dimension of M. There can be other peaks in the Patterson of the crystal X which also lie within this circle, and these do not fit the search Patterson function.

The procedure, then, is to calculate the Patterson function of the model molecule M, and to rotate it over the Patterson function of crystal X, keeping the origins at the same point. For every possible rotation, the agreement between the two Pattersons is evaluated, using only the region within a radius r of the origin. If the two agree well within this radius, I have probably found the orientation of molecule M in the crystal structure X. A function that calculates this agreement for all possible rotations is called a *rotation function*.

Rotational searches are really more complex

Box 9.1 explains how to search for a function Q which agrees well with P. Box 9.2 shows how this idea may be used in a computational search to identify the rotation which best fits one Patterson function to another. In reality, these searches have to be conducted on three-dimensional structures, with rotations described by three angles. Moreover, whatever symmetry crystal X possesses makes its Patterson function more complex. Rotational symmetry, such as a 2-fold axis, will create two or more sets of intramolecular vectors, related by the symmetry rotation, and the Patterson of the M molecule will only superimpose on one set at a time. In addition, there are vectors in the Patterson of the crystal X between an atom in one molecule and an atom in a different molecule. These vectors are called cross-vectors, and they do not match the Patterson of M. As the symmetry becomes more complex, more cross-vectors exist, but many of them lie far from the origin, not within the radius of the molecule.

Box 9.1 Searches that fit two functions

A simple method of evaluating the fit between two sets of values is to form the product of each pair. Suppose we are fitting two Patterson functions, which may be negative as often as they are positive (except near the origin where they must be positive). The product gives positive values when the two functions agree in sign (even if both are negative), and negative values when they disagree. Taking the point-by-point product of the two functions, and adding up all the products, gives a measure of their agreement.

According to this means of evaluation, writing the two functions as $P(x, y, z)$ and $Q(x, y, z)$ and fitting them over a volume V,

$$\text{Fit} = \iiint_V P(x, y, z)\, Q(x, y, z)\, \mathrm{d}x\, \mathrm{d}y\, \mathrm{d}z,$$

which may be written more briefly as:

$$\text{Fit} = \int_V P(\boldsymbol{x})\, Q(\boldsymbol{x})\, \mathrm{d}V. \tag{1}$$

A more sophisticated method is to calculate what statisticians call a *correlation coefficient*. This compares P and Q to their mean values within the volume V, and then compares the product of these to the maximum possible value. The correlation coefficient has a value exactly 1 if P and Q are identical everywhere; and it is zero if P and Q are totally independent. If $P = -Q$ throughout V, the correlation coefficient is -1.

The correlation coefficient is defined as:

$$\frac{\int_V (P(\boldsymbol{x}) - \overline{P})(Q(\boldsymbol{x}) - \overline{Q})\mathrm{d}V}{\left(\int_V (P^2(\boldsymbol{x}) - \overline{P}^2)\, \mathrm{d}V \int_V (Q^2(\boldsymbol{x}) - \overline{Q}^2)\, \mathrm{d}V\right)^{1/2}}, \tag{2}$$

where $\overline{P}$ is the mean value of P, $(1/V) \int_V P(\boldsymbol{x})\mathrm{d}V$, and similarly for $\overline{Q}$.

This slightly daunting expression is based on the same principle as eqn 1. The numerator makes a product of two quantities which will be positive if P and Q agree. The denominator scales the result to give values in the range $+1$ through 0 to -1, representing perfect correlation, no correlation at all, and perfect negative correlation.

To search for a function that fits to P, we may evaluate eqn 1 or 2 for various possible functions Q until we find the one that gives the highest value.

These factors cause the search to be not only computationally demanding, but also noisy. It is actually surprising how well the rotation function works! To reduce the noise due to cross-vectors, the maximum radius r used in fitting the Patterson functions must be kept small. It is usually best for r to be slightly less than the maximum dimension of the molecule, since this eliminates many cross-vectors, and only a few self-vectors.

The result of the rotational search also depends on the resolution range $d_{\min}$ of the intensity data used in the search. Serious noise is introduced into the function by reflections omitted from the intensity data, so data completeness is important. Often, the best results are obtained by including as much data as possible, but difficulties can arise at high

---

Box 9.2 The rotation function

The Patterson function of the model structure will be called $P_M(\mathbf{u})$ and that of an unknown structure, X, will be called $P_X(\mathbf{u})$. The symbol $\mathbf{u}$ is used to represent a point in the three-dimensional space of the Patterson function.

A model M for a single unit creates a single, non-repetitive distribution of Patterson density, while the crystal structure X is a lattice function having the lattice symmetry of the crystal (see Figs 9.6–9.8).

In calculating the rotation function, the model Patterson density $P_M(\mathbf{u})$ needs to be rotated. The value of the rotated Patterson function is written $P_M(\mathbf{Cu})$, where C is a rotation operation carried out on $\mathbf{u}$ to rotate it about the origin. ($\mathbf{Cu}$ may be called the vector obtained by multiplying the vector $\mathbf{u}$ by a rotation matrix C.)

To calculate the rotation function, the fit between $P_X(\mathbf{u})$ and $P_M(\mathbf{Cu})$ is evaluated for all possible rotations, C. Taking the simple definition of the fit given in eqn 1 of Box 9.1, the rotation function R for a particular rotation C is written:

$$R(C) = \int_{|u|<r} P_M(\mathbf{Cu})P_X(\mathbf{u})\, dV. \tag{1}$$

The integration adds up the product $P_M(\mathbf{Cu})\, P_X(\mathbf{u})$ over all values of $\mathbf{u}$ with magnitude, $|\mathbf{u}|$, less than the chosen radius, r.

It is worth remembering that the rotation operation C requires three angles to define it, and that many values of R(C) need to be calculated to explore the function thoroughly.

If eqn 1 is used, it is customary to calculate the mean value $\overline{R}$ obtained from the complete set of n computed R(C) values, and their standard deviation σ, which is the root-mean-square deviation from the mean :

$$\sigma(R) = (\Sigma_n(R(C) - \overline{R})^2/n)^{1/2}.$$

For a value of the rotation function to be highly significant, $(R(C) - \overline{R})$ must exceed 4σ. It should also be at least 20% higher than any value of R which is non-significant, that is to say one that results from random noise.

Procedures exist to refine the values obtained by a rotation function search, so as to find accurately the highest local value of the rotation function, which may not be precisely at one of the search values.

---

resolution from inaccuracy in the M structure's representation of X and from the large volume of data. When the results of a rotation search are presented, r and d_{min} should both be stated. Recent automatic procedures produce results using a variety of different values.

A rotation function will always have many peaks, some larger than others. Since it is possible for the correct peak to be obscured by noise, statistical tests are usually made. Most frequently, the mean and standard deviation of all the points in the rotation function are computed, and the significance of the highest peak is demonstrated by stating its deviation from the mean (Box 9.2). Often the magnitude of the next highest peak is stated, which must usually be assumed to represent random noise.

Choosing an incorrect peak in the rotation function will cause the attempt to interpret structure X to fail. If there is any doubt at all, it is best to keep a number of possible candidates for the rotation angle in consideration, when proceeding to step 2. To eliminate risk of error, and to maximize the chance of a successful search, some workers recommend working with the top hundred or more peaks in the rotation function, all as possible solutions to the rotation problem of step 1.

For computational reasons, it is often more efficient to search for the correct rotation using a different system of angles, known as the Eulerian angles (Fig. 9.9), rather than the (angle of rotation, colatitude of axis, longitude of axis) or (κ, ψ, ϕ) system already described. More details are given in Box 9.3.

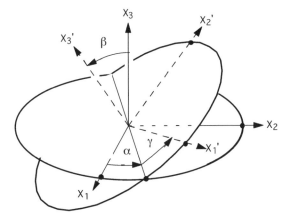

Fig. 9.9 The three perpendicular directions X_1, X_2, X_3 are converted to X_1', X_2', X_3' by rotations through the Eulerian angles α, β, γ.

Box 9.3 Systems of rotation angles

In the main text, rotations in three dimensions are described in the (κ, ψ, ϕ) system, where κ is the angle through which the body is rotated and ψ, ϕ designate the direction of the rotation axis by defining its colatitude and longitude (Fig. 9.2).

In practice, computation is more efficient in terms of a rotational system devised by Leonhard Euler, known as the *Eulerian* angles α, β, γ (Fig. 9.9). This system deals more efficiently with the equivalence of rotations which result from crystallographic symmetry. Many computer programs present the results in terms of these angles, but the rotations are more difficult to visualize, and create difficulty in presentation.

In all rotational systems, the shapes of peaks tend to be distorted. In the (κ, ψ, ϕ) system, for example, if the ψ angle is zero (meaning rotation about the 'north pole' of the sphere shown in Fig. 9.2), the ϕ value is not significant. At the north pole, all longitudes converge. Thus in a tabulation of a rotation function in terms of these angles, the values for $(\kappa, 0, \phi)$ are the same for all values of ϕ. The same is true for rotations with $\psi = 180°$, which represent rotation about the 'south pole'.

In the Eulerian system, there are many representations of a zero rotation. The rotation $\alpha, 0, -\alpha$ is a zero rotation for all angles α. The first section $(\beta = 0)$ of a printout of the rotation function contains a diagonal line all showing the same value, representing zero rotation, which may appear confusing and surprising. The section $\beta = 180°$, which may be the last section to be presented, has similar properties.

Example of a rotation function search: *Alcaligenes* esterase (Bourne *et al.* 2000)

A novel dimeric esterase of 2×35 kDa was extracted from the soil bacterium *Alcaligenes*, and crystallized in unrelated orthorhombic and monoclinic crystal forms; the monoclinic crystals included a substrate analogue. The orthorhombic crystal structure was determined by MIR at 2.7 Å resolution to give a preliminary structural model.

To study the monoclinic form, the first step was a rotation function, one section of which is shown in Fig. 9.10. Its peak, which exceeds 7 standard deviations, and is far higher than any other, suggested an orientation of the search model which could be used in further structure analysis. The Eulerian angle rotation system was used in the search (Box 9.3).

Translation searches to derive a model structure

Step 2 is to decide where the origin of the model molecule M must be placed, to generate a crystal structure with diffraction corresponding closely to that of X. This step uses a *translation function*, which searches for the three positional parameters when the three rotational parameters have been determined. Once again, it uses the Patterson function, so that phase information is not needed. In essence, the search consists in placing the origin of M at all positions in the unit cell of X. For each position, the predicted configuration of Patterson vectors is compared to the actual Patterson function of X. Box 9.4 explains a two-dimensional example, and Box 9.5 outlines how the method is applied in practical cases.

Compared to the rotation function, a translation function can be computed very rapidly. If there is any doubt about the correct rotation, it is therefore reasonable to repeat the translation search for a number of candidate rotations. There is still a huge economy by

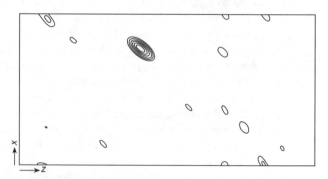

Fig. 9.10 One section ($\beta = 55°$) from the rotation function plot searching the Patterson of the monoclinic form of the esterase, using a preliminary esterase model built into 2.7 Å MIR electron density from the orthorhombic crystal form as search model. The search was made over an 18 Å sphere around the origin of the Patterson, using data between 10 and 4 Å resolution. The quantities labelled x and z are the Eulerian angles α and γ. The contour levels represent increments of one standard deviation (σ) of the rotation function, with the lowest contour at 3σ. (Illustration kindly provided by Micha Isupov and Jenny Littlechild.)

Box 9.4 Translation function in two dimensions assuming 2-fold symmetry

From the result of a rotation study, the model molecule M is rotated so that it lies parallel to one of the molecules in crystal X, as in Fig. 9.5.

Now, note that the problem of positioning a molecule in a unit cell only exists if there is some crystal symmetry. If the crystal has no symmetry other than the lattice symmetry, the origin of the crystal lattice can simply be defined as that point corresponding to the origin of the model. But if there is rotational symmetry in crystal X, the crystal origin must be defined relative to the position of its rotation axes. The problem is, therefore, to find the position of rotational symmetry axes in the crystal relative to the origin of the model coordinates.

As an example, assume X has 2-fold symmetry, and imagine step 1 has been completed successfully. So, we already know how to rotate the model M to make it parallel to a molecule in X. Now, make a dimer model using this rotated M and a copy M' rotated by a 2-fold symmetry axis about any chosen origin (Fig. 9.11). The Patterson function of this set of atoms includes two kinds of vectors, self-vectors and cross-vectors (Fig. 9.12). As there are six atoms in MM', the origin peak has weight 6. There are 12 self-vectors (6 double-weight) and 18 cross-vectors (6 double-weight and 6 single-weight).

In the unknown structure X, the molecule is rotated about a parallel axis, but it is in an unknown position (Fig. 9.13). In its Patterson function (Fig. 9.14) the self-vectors are unchanged, but the cross-vectors are displaced. Although the cross-vectors are displaced, the shape of the constellation of cross-vectors is unchanged.

The purpose of the translation function is to determine the vector between the origin assumed in the model MM' and the position of the actual origin in the unknown structure X (Fig. 9.15).

The constellation of cross-vectors has been generated from MM'. Imagine sliding it over the Patterson function of the crystal X, until it fits onto the same constellation of cross-vectors in the crystal Patterson function. The vector through which the constellation has been displaced, from its starting position to the position of best fit (Fig. 9.16), determines the position of the model's origin relative to the position of the 2-fold axis in X (Fig. 9.15).

The translation function evaluates the fit of the constellation of cross-vectors from the model to the experimentally derived Patterson function for all possible displacement vectors. The translation function should have a large value for one particular displacement vector.

comparison with a full six-dimensional search of all possible rotations and translations. In this way, the correct solution may be revealed, even though the rotation search was partly masked by noise.

It is obvious that for any chosen rotation, some positions of the origin of M are ruled out by packing considerations. The centre of the molecule cannot be placed close to a rotation axis of symmetry, for example. If M is placed at such a position, a symmetry-related molecule will overlap with it. When a large peak is found in a translation function, it is reassuring to check that the indicated molecular position causes no serious steric conflict.

Fig. 9.11 The model M and a 2-fold rotated copy M'. The 2-fold axis position is shown.

Fig. 9.12 The Patterson function of the dimer MM'. Self-vectors between atoms of a monomer are filled, while cross-vectors between M and M' are open circles. The cross-vectors from M' to M are outlined by a triangle.

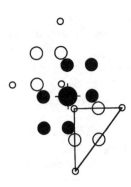

Fig. 9.13 The unknown structure X, composed of molecules similar to M and M', related by a 2-fold axis in a different position.

Fig. 9.14 The Patterson function of the dimer X. The constellation of self-vectors (filled circles) is identical to that of the model MM'. The cross-vectors are in different positions, but the constellation of cross-vectors picked out by the triangle is identical to that picked out in the Patterson of MM'.

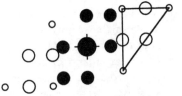

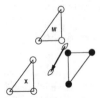

Fig. 9.15 The structures MM' and X are overlaid, so that one molecule coincides. The arrow shows the vector from the 2-fold axis assumed for the model to its position in the unknown structure X.

Fig. 9.16 The Patterson functions of MM' and X are overlaid, using the same origin. The self-vectors are identical (black circles). The cross-Patterson vectors of X are shown as clear circles; those of MM' are shaded. The vector relating the two constellations is double the vector that relates the position of the 2-fold axes in MM' and in X.

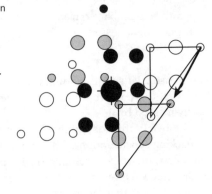

Box 9.5 Translation function in three dimensions

The two-dimensional example of Box 9.3 is particularly simple, because in two dimensions a 2-fold rotation is precisely the same as an inversion operation—the 2-fold symmetrical dimer has a centre of symmetry in two dimensions. The Patterson function always has a centre of symmetry, and in this particular case the 2-fold symmetry does not add any further features.

Three-dimensional protein or nucleic acid models never have a centre of symmetry. But if the example shown in Figs 9.10–9.15 is reworked in three dimensions, with each atom of the model M at a different height above the paper, the overlap of the shifted model Patterson and the Patterson of the dimer works as before.

A translation function can be used to find the axial position of any type of rotational symmetry operation, including screw axes. In operations of higher-order symmetry, the translation function still compares the cross-vectors between two particular molecules.

Just as in the simpler case, there is a constellation of cross-vectors which can be precisely overlapped. The displacement of the origins between the model and the unknown structure may be defined by a vector $\mathbf{t}$. The displacement of the constellation of Patterson cross-vectors in this more general case can be written as $\mathbf{t} - R\mathbf{t}$, where $R\mathbf{t}$ represents the vector $\mathbf{t}$ operated on by the rotation R relating the subunits being compared. For a 2-fold axis in two dimensions, $R\mathbf{t} = -\mathbf{t}$, which explains why in Fig. 9.16 the displacement $\mathbf{t} - R\mathbf{t} = 2\mathbf{t}$.

In the majority of cases, a correct peak in a translation function is overwhelmingly significant. It is, however, desirable to compare it to the standard deviation from the mean value of the translation function, and to make sure that it represents well over 4 standard deviations.

Example: translation function search for *Alcaligenes* esterase (Bourne *et al.* 2000)

When the search model for the *Alcaligenes* esterase was oriented as indicated by the rotation function peak in Fig. 9.10, a translation function gave an overwhelming peak, indicating the placement of the model molecule. Since the monoclinic crystals gave measurable diffraction to 1.1 Å resolution, this allowed the structure to be determined at much higher resolution than was possible for the orthorhombic form.

Using a model structure to generate phase angles

The purpose of all this searching is to generate an approximate electron density for the unknown structure X, which can be done from the model M when the correct rotation and translation have been identified. To explain the ideas, we originally assumed that the molecules in M and X were identical, but it will now be evident that the same procedures can work pretty well when they have some differences. In an extreme case, X might be a molecule which is only weakly similar to M; or X might be a complex in which M represents a minor part.

The diffraction intensities of X have been measured, but to generate the crystal structure phases have to be estimated for each reflection. This estimate is made by calculating the phase angles implied by the model structure appropriately oriented in the cell of the unknown structure.

At this stage, the phase angles $(\alpha_{hkl})_{\mathrm{calc}}$ generated from the calculated scattering of the model structure are the best estimate available. To make the best estimate of electron density, they must be attached to the observed structure amplitudes $|F_{hkl}|_{\mathrm{obs}}$ (Box 9.6).

Box 9.6 How to get an electron-density map from molecular replacement results

When the rotation and translation operations are known, these can be applied to all the atoms of the model M, placing them at appropriate positions in the crystal structure. The position $\mathbf{x}_M$ of an atom in M is transformed to a position $\mathbf{x}_X = \mathbf{R}\mathbf{x}_M + \mathbf{t}$ in the crystal structure X, where R is a rotation operation and $\mathbf{t}$ a translation vector.

The set of $\mathbf{x}_X$ values will represent the structure of one M-like molecule, appropriately placed in the unit cell of X.

Estimated structure factors of the crystal structure can then be calculated according to eqn 2 of Box 5.2. As stated there, the summation has to be taken over all the N atoms in the unit cell. This means including not only the atoms at $\mathbf{x}_X$ as described above, but the atoms from other units generated by the crystal symmetry from the subunit X in the crystal.

In this way a set of calculated structure factors F_{calc} may be obtained :

$$F_{\mathrm{calc}}(\boldsymbol{h}) = |F_{\mathrm{calc}}(\boldsymbol{h})|\exp[i\alpha_{\mathrm{calc}}(\boldsymbol{h})]$$
$$= \sum_N f_i \exp[2\pi i \boldsymbol{h}\cdot\boldsymbol{x}_i]\,\exp[-4B_i\sin^2\theta/\lambda^2]. \tag{1}$$

Unless further information is available from isomorphous replacement or some other method, $\alpha_{\mathrm{calc}}(\boldsymbol{h})$ is the best information about the phase angle $\alpha(\boldsymbol{h})$. But experimental information exists for $|F(\boldsymbol{h})|$, namely the measured structure amplitude $|F_{\mathrm{obs}}(\boldsymbol{h})|$. Therefore the best estimate for $F(\boldsymbol{h})$ is given by

$$F(\boldsymbol{h}) = |F_{\mathrm{obs}}(\boldsymbol{h})|\,\exp[i\alpha_{\mathrm{calc}}(\boldsymbol{h})]. \tag{2}$$

An electron-density map for $\rho(x)$ is therefore calculated (using eqn 3 of Box 5.2) with Fourier coefficients $F(\boldsymbol{h})$ indicated in eqn 2 above.

The use of calculated phase angles α_{calc} in this electron-density map means that it does not present an unbiased estimate of the electron density. It will tend to be biased towards the model structure used to generate the phase angles. A feature of the model M which does not represent the structure X correctly is still likely to appear in the electron-density map, but more weakly than in the starting model. Equally, any atoms in X which do not correspond to an atom in M will only appear weakly in this map.

It is particularly important to deal correctly with regions of X not covered by the model structure M, and to allow for the electron density of solvent which fills empty parts of a macromolecular crystal.

Improvement of a molecular replacement model

A model for the crystal structure X has now been generated by building it up from appropriately oriented molecules identical to M. This model needs to be improved. The structure factors calculated from the model will have magnitudes $|F_{\text{calc}}|$ which differ from the observed structure amplitudes $|F_{\text{obs}}|$. The differences between these can be used to calculate a residual map, which should indicate how the electron density needs to be changed (Box 9.7). Features of X not included in the model appear as positive density, and features of the model that do not fit the real density should appear as negative density. Careful study of this difference map will allow the proposed structure to be made more exact, compared to a precise copy of the model M. This process can be assisted by computational procedures to be discussed in Chapters 11 and 12.

Application of molecular replacement to complexes like enzyme-substrate or enzyme-inhibitor complexes

Often X is expected to have specific features different from M. A typical example would be where X is an enzyme-inhibitor complex, while M models only the enzyme. A residual map can give excellent quality electron density for a small ligand molecule, which can often be interpreted to give atomic positions.

Box 9.7 Electron-density difference map to look for errors in the starting model

Box 9.6 has explained how the electron-density map generated by molecular replacement may be calculated, but has shown how it is likely to be biased towards the features of the model. To study this bias, it is useful to compute an electron-density difference map. This shows the difference of electron density between the starting model and that obtained from molecular replacement.

The $F_{\text{calc}}(\boldsymbol{h})$ of eqn 1, Box 9.6 represents the structure factors of a crystal built up from copies of the starting model M. The best estimate for $F(\boldsymbol{h})$ is given by eqn 2 of Box 9.6, using observed structure amplitudes with phase angles calculated from the model. The difference between these is

$$F(\boldsymbol{h}) - F_{\text{calc}}(\boldsymbol{h}) = |F_{\text{obs}}(\boldsymbol{h})| \exp[i\alpha_{\text{calc}}(\boldsymbol{h})] - |F_{\text{calc}}(\boldsymbol{h})| \exp[i\alpha_{\text{calc}}(\boldsymbol{h})]$$

$$= (|F_{\text{obs}}(\boldsymbol{h})| - |F_{\text{calc}}(\boldsymbol{h})|) \exp[i\alpha_{\text{calc}}(\boldsymbol{h})].$$

Use of these coefficients in a Fourier transformation creates the difference map:

$$\rho_{\text{diff}}(\boldsymbol{x}) = \frac{1}{V}\sum_{h}(|F_{\text{obs}}(\boldsymbol{h})| - |F_{\text{calc}}(\boldsymbol{h})|) \exp[i\alpha_{\text{calc}}(\boldsymbol{h})] \exp[-2\pi i\boldsymbol{h}\cdot\boldsymbol{x}].$$

This difference map, an '$(F_o - F_c)$ map', will have positive features representing density present in the crystal structure which are not in the model. Its negative features represent density created by the model which is not in the crystal structure.

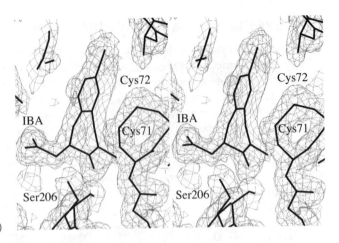

Fig. 9.17 Electron density near the active site of the esterase, showing density not accounted for by the protein model, interpreted as the IBA substrate analogue. (The map shown here is a $(2F_o - F_c)$ map, to be described in Chapter 11.) (Figure reproduced from Bourne *et al.* (2000), with permission from Elsevier Science.)

Example: density for substrate analogue of *Alcaligenes* esterase (Bourne *et al.* 2000)

As mentioned above, the monoclinic crystal form of the *Alcaligenes* esterase included a substrate analogue, IBA. When the electron-density map was computed using phases calculated from the refined model based on the orthorhombic form, density for the IBA molecule was clearly presented (Fig. 9.17).

Further reading

A recent survey of molecular replacement can be found in:
Rossmann, M.G. and Arnold, E. (2001). Patterson and molecular-replacement techniques. In *International tables for crystallography*, Vol. B (ed. U. Shmueli), pp. 235–63. International Union of Crystallography/ Kluwer, Dordrecht.

More detailed material, including sections by D.M. Blow, J. Navaza, and L. Tong will be found in:
Rossmann, M.G. and Arnold, E. (ed.) (2001). Molecular replacement. In *International tables for crystallography*, Vol. F, Chapter 13, pp. 235–63. International Union of Crystallography/Kluwer, Dordrecht.

Another useful survey, concentrating more on computational systems, including articles by A.T. Brünger, J. Navaza and P. Saludan; L. Tong and M.G. Rossmann; and G.A. Bentley is in:
Carter, C.W. and Sweet, R.M. (ed.) (1997). Macromolecular crystallography. IV. C. Molecular replacement. *Methods Enzymol.*, **276**, 558–611.

10

Density-modification procedures: improving a calculated map before interpretation

Procedures to determine the phases of the structure factors, by isomorphous replacement, by anomalous scattering, or by molecular replacement, were described in the Chapters 7–9. Using one or more of these methods, phases are generated which allow an electron-density map to be calculated, at a resolution to which the phases are thought to be reliable. In many cases this electron density can be confidently interpreted in terms of atomic positions. But this is not always the case.

Quite often, the procedures so far described offer a tantalizing puzzle map, with some features which I think I can interpret, but raising many questions. Before devoting effort to interpreting an unsatisfactory electron-density map, a number of procedures are available, which might make a striking improvement.

Perhaps the most important strategy is to seek out more isomorphous and anomalous scattering derivatives. Before doing that, there are other possibilities which may improve an electron-density map without any more experimental data. These methods are known collectively as *density modification*.

Improvements based on solvent structure

The first group of methods exploits features of the electron density which result from the packing of molecules into a crystal. Macromolecular crystals composed of rigid molecules have voids between the molecules filled with disordered solvent, often including the precipitants used in the crystallization process. These solvent regions present featureless density between the structured density of the macromolecules.

A high-quality electron-density map will show these featureless regions clearly. In a map of poorer quality, the voids between molecules may be clearly defined, but far from featureless. This provides a method to improve the map.

Although some solvent molecules are immobilized on the surface of the macromolecule, those further from the surface are in a disordered liquid-like state which presents a uniform density. Except in very small proteins, the majority of solvent is disordered.

If such uniform solvent regions can be recognized, they allow surfaces to be defined which separate solvent regions from protein regions.

Two procedures are described below. It has become almost a matter of routine to use one or both of these methods.

Solvent flattening

In the *solvent-flattening method* the electron density of solvent regions is modified and set to a mean value, while the protein part of the map is unaltered. The resulting electron-density distribution is used to calculate structure factors, the phases of which are considered to be an improvement on the starting phases. A 'combined' set of phase angles may be created, forming a compromise between the experimental phases and the phases calculated from the modified electron density. These phase angles are used with the observed structure factors to calculate an 'improved' electron-density map. This map will show much flatter density in the regions assigned as solvent. If the solvent regions still show variations, the procedure may be repeated cyclically.

This procedure has been extremely successful in improving many electron-density maps. Although the labour of defining the molecular surface has been minimized by computer techniques, there is still some arbitrariness in deciding which parts of a map represent solvent. A serious hazard is that parts of the structure might be assigned as solvent. If this happens, the procedure will do its best to wipe them out. Therefore it is most important to assign solvent regions conservatively, and to allow generous boundaries around the molecules.

Example: galactose oxidase (Ito 1991)

In Chapter 8, improvements to the phase angles of galactose oxidase by SIRAS and MIRAS procedures were presented. Figure 8.15 shows the further improvement to the phase angles obtained by solvent flattening. The phase improvement is particularly important at resolution beyond about 3.2 Å, where MIRAS phasing is becoming poor.

Histogram matching

A related procedure called *histogram matching* may be applied to the protein part of a protein electron-density map. To explain it, let's begin by explaining a histogram. Suppose we are looking at all the electron-density values within a protein molecule. We can count how many times the electron density is between 0 and 1 unit, and plot this number in position 1, how many values are between 1 and 2 units, and plot this at position 2, and continue until we have counted the number of times the highest recorded density was observed. We can do the same for any negative densities that appear. Such a histogram is shown in Fig. 10.1, compared with a histogram calculated from the refined structure.

The histogram will cover the whole range of values that occur in the electron density of the protein. The electron-density features of proteins are now very well known. Given the resolution, the overall B factor, the overall scale, and other parameters, the expected form of the histogram of electron-density values can be predicted with precision. Deviations from the predicted histogram must result from incorrect phasing. Errors in the phases determined experimentally tend to smear out the histogram, giving a broader distribution than the true one.

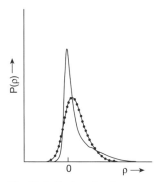

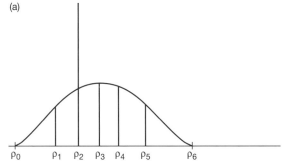

Fig. 10.1 Electron-density histograms of pig insulin at 1.9 Å resolution. ⚫–⚫ Histogram from a map calculated from multiple isomorphous replacement phases. — Histogram from a map using phases calculated from refined coordinates. (Reproduced from Zhang and Main (1990), by permission of the International Union of Crystallography.)

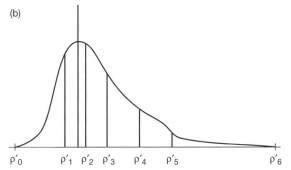

Fig. 10.2 An experimental histogram (a) and the expected histogram (b) have each been divided into six equal areas. To generate the expected histogram, each electron-density value has to be modified. An electron density at the level marked as ρ_i in (a) must be converted to the value ρ'_i in (b). (Reproduced from Woolfson and Fan (1995), by permission of Cambridge University Press.)

The *histogram matching* procedure is to reassign every electron density within the protein volume to a value consistent with the predicted distribution of densities for the structure, when observed at this resolution (Fig. 10.2). Thus, the maximum and minimum calculated densities are adjusted to the range that can be found in a protein image at this resolution. Intermediate densities are adjusted so that the predicted histogram of electron densities is obtained. In this way, a modified electron-density map may be prepared which should be a truer representation of the structure. It can be used to calculate 'improved' phase angles, which in turn produce an 'improved' map. This whole procedure may be recycled if desired.

Prediction of the histogram for structures containing extensive non-protein regions, such as a protein-nucleic acid complex, would raise further difficulties, and this method is normally used only on proteins.

Example: hCASK PDZ domain (Daniels *et al.* 1998)

The 93-amino-acid PDZ domain from the protein hCASK was analysed to 2.7 Å resolution. MAD phasing from crystals of selenomethionine-substituted protein gave a mean figure of merit of 0.27. A 'representative' part of the resulting electron-density map is

(a)

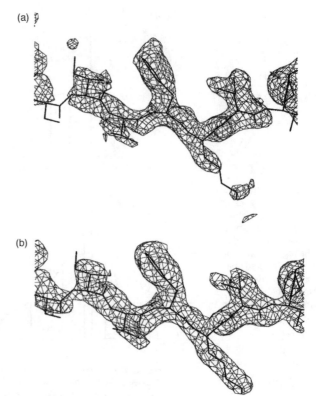

(b)

Fig. 10.3 (a) Electron density at 2.7 Å resolution before solvent-flattening and histogram-matching procedures. (b) The same electron density after 2-fold symmetry averaging, solvent flattening, and histogram matching. (Reproduced from Daniels *et al.* (1998), by permission of Nature Publishing Group.)

compared in Fig. 10.3 with the same part of the map after 2-fold symmetry averaging, solvent flattening, and histogram matching. It may be observed that the break in main-chain electron density is partly corrected, and the density of several side chains is considerably improved.

More than one subunit in the crystal asymmetric unit

A further group of methods exploits the extra information available when more than one copy of a structure is found within the asymmetric unit of a crystal. Methods to exploit this situation are closely related to the methods for molecular replacement described in Chapter 9.

Often a protein molecule is an oligomer. For generality, the word 'subunit' will now be introduced. The subunit might very well be a molecule, but it is often a monomer unit of an oligomer. It could even be a smaller unit. Or, it could be a group of several molecules forming part of a larger complex.

Suppose the repeating unit of the crystal contains more than one identical object. Stated in more technical language, suppose there is more than one subunit in the crystal asymmetric unit.

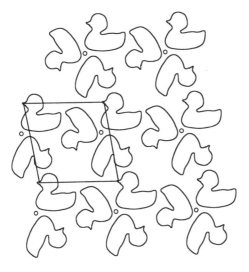

Fig. 10.4 The repeating unit of this crystal contains three subunits. Groups of three subunits are related by a 3-fold symmetry axis, but this 3-fold symmetry does not apply to the whole crystal.

When molecules consist of oligomeric units in solution, these oligomers are usually symmetrical. The individual monomers are usually related by one of the point-group symmetries (Fig. 2.18). But often the oligomeric molecule crystallizes with the oligomer as the repeating unit. The internal symmetry of the oligomer does not show up in the crystal symmetry. A simple two-dimensional example is shown in Fig. 10.4. The trimeric subunits are arranged symmetrically, but this 3-fold symmetry does not apply to the crystal as a whole. Since the symmetry applies locally, but not throughout the crystal, this situation is called *local* or *non-crystallographic symmetry*.

Non-crystallographic symmetry is surprisingly common. In this situation the crystallographer has more information than usual. The number of measurable intensities depends on the size of the asymmetric unit (Box 5.6), but the unknown subunit is smaller than the crystal asymmetric unit. If there are three subunits in the crystal asymmetric unit, the structure will provide about three times as many intensity values for measurement.

It is hard to use this method to determine new structures directly, but it has proved powerful in improving poorly interpretable electron density, and occasionally in cautiously extending resolution limits. Its triumph has been to make structure determination for the capsids of spherical viruses a routine procedure.

The first step is to find out how the subunits are arranged in the crystal structure. If the subunits are rigid, one subunit is related to another by an operation of rotation and translation. The problem is similar to that of molecular replacement (Chapter 9).

The self-rotation function and its interpretation

To determine the rotation of one subunit to another in the same structure, a rotation function may be used which rotates the Patterson function upon itself, a procedure called the 'self-rotation function' (Box 10.1). This evaluates the agreement of the Patterson with a rotated version of itself, over a volume V near the origin, for all possible rotations. It should find the rotational operations that align subunits to a similar orientation.

Box 10.1 The self-rotation function

Box 9.2 shows how the rotation function can be used to evaluate the fit between the Patterson functions of a model structure M and an unknown crystal structure X, as M is rotated into all possible orientations. The self-rotation function uses exactly the same idea, to search for rotations that cause the Patterson of X to agree with itself. In both cases, matching occurs only over self-vectors of the molecule or subunit, so the search is limited to a spherical volume, V, around the origin.

A self-rotation function for all possible rotation operations C may be calculated as

$$R(C) = \int_V P_X(Cu)\, P_X(u)\, dV. \tag{1}$$

The only difference between this definition and that of eqn 1, Box 9.2 is that the Patterson function of X is being compared with a rotated version of itself, rather than the Patterson of a molecule believed to be similar to it.

To see how it works, look at two very simple molecules related by an arbitrary displacement (Fig. 10.5). Their Patterson function is shown in Fig. 10.6. Figure 10.7 shows this Patterson rotated through almost (not quite) the rotation angle that relates them. These two Patterson functions are overlaid in Fig. 10.8, where it is seen that many of the Patterson vectors almost overlap (not quite). When the rotation is exactly correct, there will be a good fit between the parts of the Patterson functions near to the origin.

These diagrams have been kept more simple by considering only isolated molecules. In a real case, the crystal lattice generates many more cross-vectors, but the principle remains the same.

In Fig. 10.8, the overlapping parts of the two Patterson functions are all within a radius corresponding to the maximum dimension of a molecule. To maximize the amount of overlap, while excluding the cross-vectors, none of which overlap, the volume V should include density round the origin to no more than this radius.

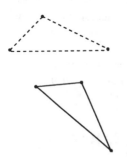

Fig. 10.5 A dimer, the two subunits of which are related by arbitrary rotation and translation. One member of the dimer is indicated by dashed lines joining its atoms.

A simple two-dimensional example is shown in Fig. 10.5, choosing a subunit of three atoms. A non-symmetric dimer is shown, in which the subunits are related by an arbitrary rotation and translation. Figure 10.6 shows the Patterson function of the dimer. Figures 10.7 and 10.8 show how a rotated version of this Patterson can make a good fit to the unrotated Patterson.

As a copy of the Patterson function is rotated over the original, one can search for rotations where the two Patterson functions agree with each other. The self-rotation function, just like the rotation function described in Chapter 9, evaluates how well the rotated Patterson fits to the original, for all possible rotations. A rotation that creates a good fit indicates the actual rotation needed to align one subunit to the other.

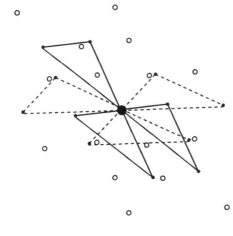

Fig. 10.6 The vector set representing the Patterson function of this dimer. The intra-subunit vectors of each subunit are indicated by filled circles linked by lines; the inter-subunit vectors are shown as open circles.

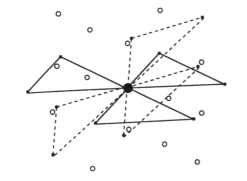

Fig. 10.7 The Patterson of the dimer has been rotated so that the self-vectors, with full lines, have become almost parallel to those with dashed lines in the original Patterson function.

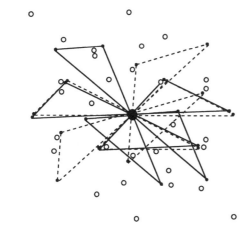

Fig. 10.8 In this diagram, the self-vectors of the original Patterson are shown with full lines. The rotated version of the Patterson (dashed lines) has been superimposed on it. If the rotation were exactly correct, half of the intra-subunit vectors of one Patterson would super-impose on to each other. None of the inter-subunit vectors would superimpose.

Box 10.2 Features of the self-rotation function

Self-rotation functions always have a huge value for zero rotation. Obviously, a Patterson function will agree perfectly with a copy that has not been rotated—that is, rotated through a zero angle. So this tells you nothing, and should be ignored. When the rotation function is plotted as a function of the rotation angles, this high value creates a peak referred to as the 'origin peak'.

Any rotation that represents a crystallographic symmetry operation will have an equally huge peak. Thus, if a crystal has 2-fold rotational symmetry, this 2-fold rotation operation will also create a copy of the Patterson which is identical to the original. In the calculated rotation function, peaks equal to the origin peak will be found at every rotation representing a crystallographic symmetry operation.

In order to review all possible rotations, three angles have to be varied, as already discussed in Box 9.3. In the (κ,ψ,ϕ) system, if the rotation angle κ is zero, the rotation is zero, whatever the values of ϕ and ψ. If you review a plot of a self-rotation function as a three-dimensional plot in terms of these angles, the whole section of the plot that has $\kappa = 0$, covering all values of ϕ and ψ, represents the origin peak.

As mentioned in Box 9.3, zero rotation is expressed by Eulerian angles with $\beta = 0$, wherever $\alpha = -\gamma$. The result is that the $\beta = 0$ section of a self-rotation function is covered with a swathe of very large values, all representing zero rotation.

The huge origin and crystallographic symmetry peaks of the self-rotation function must be ignored. We must search for a smaller peak or peaks which are the solution to the rotation problem. As before, these peaks must be substantially larger than any other peaks which do not represent self-rotation operations, and several standard deviations more than the 'noise' or meaningless peaks. (But, in estimating this standard deviation of the function, it is permissible to ignore the very large values found around the origin and symmetry peaks.)

In three dimensions, the self-rotation function can be plotted in terms of the three angles that describe any rotation, and there should be a large peak at a rotation that represents the rotation of any subunit to align it with another. More detail is given in Box 10.2.

Example of a self-rotation function search: D-glycerate dehydrogenase (Goldberg 1993)

D-glycerate dehydrogenase crystallizes with two molecules in the crystal asymmetric unit. The crystals have no symmetry other than the lattice translations (space group P1), so this means there are two molecules in the unit cell. Since the crystal symmetry has no rotation operations, the self-rotation function is unusually simple to interpret.

Figure 10.9 shows the self-rotation function for rotations of 180° about all possible directions. The directions are shown schematically as directions on the surface of a sphere. The rotation function is plotted as a contour map. One rotation gives a large value of for the rotation function—more than 6 standard deviations. This is by far the highest value for any rotation (except around the origin peak).

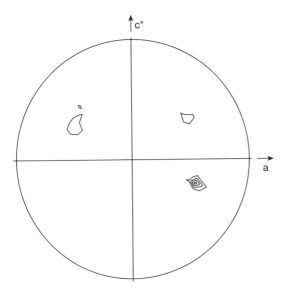

Fig. 10.9 The κ = 180° section of the self-rotation function for triclinic D-glycerate dehydrogenase crystals. The lowest contour encloses regions where the self-rotation function exceeds 3 standard deviations. Higher contours are drawn at 4, 5 and 6 standard deviations. (Reproduced by permission from Goldberg 1993.)

Positioning the subunits

The arrangement of the subunits now has to be identified. A self-translation function based on the Patterson function may be used, similar to that described in Chapter 9. Alternatively, if some phase angles are available, a poorly phased or low-resolution electron-density map can be used directly. Often the arrangement of heavy atoms introduced for isomorphous replacement gives clear evidence. In a symmetric oligomer (say a tetramer with 222 symmetry), the translation problem reduces to finding the position of the molecular centre, and the answer may even be obvious, especially for a virus molecule with 532 symmetry.

Once an approximate description of the local symmetry can be made, computer procedures are available to refine the estimated rotations and translations to more accurate values.

Exploitation of local symmetry

When this has been done, rotations and translations are known which relate a point in one subunit to one or more equivalent points in other subunits. If the subunits are rigid and identical, the electron densities at all related points should be identical. This provides a way of improving the accuracy of experimentally derived electron density.

If there are several copies of the subunit in a poorly phased electron-density map, they can be forced to be equal. The density at some particular point in each subunit may be replaced by the average density at all the equivalent points in all the subunits. When this has been done for every point in the subunit, the density distribution of the whole subunit has been replaced by an 'average' distribution derived from all the subunits. This process is often called *molecular averaging*. In the averaged density distribution errors tend to cancel out, and it gives a more accurate representation of the subunit density.

It is, however, extremely important that this averaging is only done within the volume of a subunit. If too large a volume is chosen, bits of other subunits may be averaged, introducing serious error. A carefully defined volume must be used for the averaging operation, and if any doubts exist, it is important that this volume is chosen conservatively.

In many cases there are just two subunits in the crystal asymmetric unit, and replacing pairs of electron densities by their average can already cause a significant improvement in electron density. If averaging can be done over a larger number of subunits, the improvement of the electron density can be spectacular.

Example: D-glycerate dehydrogenase (Goldberg 1993)

In Chapter 7 D-glycerate dehydrogenase provided an example where the weak phasing power of seven isomorphous derivatives was inadequate to produce an interpretable electron-density map. The self-rotation function (Fig. 10.9) located a rotation which superimposes one of the two subunits in the asymmetric unit on the other.

The fact that the highest peak represents a rotation of exactly 180° suggests that the molecule has crystallized as a symmetrical dimer. Study of the heavy-atom sites determined for the seven isomorphous replacements revealed several pairs of sites that could be related by the same 180° rotation. This allowed the position of the molecular 2-fold axis to be determined, relative to the heavy-atom sites.

An approximate envelope was determined for the dimeric molecule from a poorly phased map derived from isomorphous replacement and solvent flattening. The molecular 2-fold axis implies that the electron density at each point within the envelope should be the same at another point related by the 2-fold symmetry. The density at each pair of points throughout the whole envelope was replaced by the average of the two. This averaged electron density was used to calculate new phases, and these phases were combined with the observed structure amplitudes to compute a new map. The new map allowed an improved molecular envelope to be chosen. This whole procedure was repeated three times, by which time many interpretable features could be seen in the electron density.

Structure determination of small spherical viruses

Small spherical viruses economize genetic coding capacity by making their coat protein a highly symmetrical oligomer. The spherical coat usually has the symmetry of an icosahedron (discussed in Chapter 2). The 3-fold and 2-fold axes are arranged as in cubic symmetry, but the 5-fold axes cannot be expressed as part of the crystal symmetry. The virus coat assembly may have 60 subunits, arranged as in Fig. 10.10, but frequently small viruses have 180 subunits (Fig. 10.11).

An infective spherical virus particle is crammed with nucleic acid, but the nucleic acid is not packed with the symmetry of the coat protein subunits. The nucleic acid contributes to the crystal diffraction like a disordered liquid.

Example: tomato bushy stunt mosaic virus (Harrison *et al.* 1978)

A classic example is the virus particle of tomato bushy stunt mosaic virus, the coat of which contains 180 protein molecules, each of 43 000 Da. The crystal unit cell has body-centred cubic symmetry, containing two virus particles, each composed of 180 protein subunits, with a cubic unit cell dimension of 383 Å.

Fig. 10.10 Subunit arrangement for a virus capsid that has 60 subunits. Each of the 20 triangular faces of an icosahedron corresponds to three subunits that associate with 5-fold symmetry about each vertex of the icosahedron. (Reproduced from Harrison (1980), by permission of the Biophysical Society.)

Fig. 10.11 If there are 180 subunits in a virus capsid, three subunits, A, B, and C, take the place of one subunit of Fig. 10.10. Although they are not related by exact 3-fold symmetry, all make very similar interactions with other subunits. (Reproduced from Harrison (1980), by permission of the Biophysical Society.)

The highly symmetrical cubic unit cell contains 24 equivalent positions—that means, the crystal asymmetric unit is 1/24 of the body-centred unit cell. But this asymmetric unit contains 15 protein subunits. The number of intensities measured at any resolution is 15 times the number you would normally measure for a protein of 43 kDa.

In this case, the symmetry allows no uncertainty about the arrangement of the subunits. The body-centred cubic symmetry requires that virus particles are centred at the origin of the unit cell, and at the equivalent point at its centre. The icosahedral symmetry requires 2-fold axes to lie along the cubic cell axes, and the 3-fold axes are parallel to the body diagonals of the cell. The position of every 5-fold axis is fixed by the symmetry.

The 198 000 reflections to 2.9 Å resolution were first phased by isomorphous replacement with two derivatives, giving a mean phase error of 53° (when compared to a model where local symmetry had been enforced). Four cycles of the icosahedral averaging procedure reduced this to about 5°.

Example: rhinovirus HRV14 (Rossmann *et al.* 1985)

The coat of the common cold virus is composed of 60 subunits, each containing four different protein molecules. The subunit has a molecular mass of about 94 000 Da. The HRV14 strain crystallizes in a primitive cubic space group ($P2_13$) so that each virion lies on crystallographic 3-fold axes at the origin of the unit cell. There are 20 subunits in the crystal asymmetric unit.

Only two isomorphous derivatives were found, and these were closely similar. They used auricyanide $Au(CN)_2^-$ as heavy atom substituent at different concentrations. The changes to the diffraction pattern (expressed by R_{deriv}) were rather small (0.12 and 0.08), probably because of the large size of the subunit. Isomorphous replacement phasing was undertaken to 6 Å resolution.

Figure 10.12 displays an R factor which shows how well the non-crystallographic symmetry is obeyed (Box 10.3). The isomorphous replacement results (open circles) fitted the symmetry rather poorly.

For every point in the envelope of the rhinovirus capsid, there are 20 points in the crystal asymmetric unit which should have the same electron density, because they are equivalent points of different subunits. Averaging these 20 electron densities, and re-inserting the averaged value in every subunit, imposes the symmetry which has been determined experimentally, forcing every subunit to be identical.

The structure now has the proper symmetry, but will not agree exactly with the observed diffraction intensities. The new electron-density distribution can be used to calculate a new set of phases, α_{calc}, which are attached to the observed $|F_{obs}|$ to calculate a new electron-density distribution, that conforms more closely to the required symmetry. When there are many subunits in the crystal asymmetric unit this process converges rapidly to exact symmetry.

Two cycles of this process allowed a third, weakly occupied, heavy atom site to be identified in the rhinovirus, and improved isomorphous replacement phases could be calculated to 5.5 Å resolution. The filled circles of Fig. 10.12 show spectacular improvement in the agreement of electron density between the 20 subunits to 5.5 Å, decaying rapidly at higher resolution.

Fig. 10.12 Accuracy of non-crystallographic symmetry for rhinovirus HRV14 at various stages of the structure analysis. The data points give the agreement between observed structure amplitudes and structure amplitudes calculated from the electron-density distribution after symmetry averaging. In the upper curves (a), these are expressed as an R factor (Box 10.3). The lower curves (b) show the correlation of the observed and calculated structure factors (Boxes 10.3 and 9.2). Open circles, isomorphous replacement phases to 5 Å resolution. Filled circles, phases obtained after imposing non-crystallographic symmetry at 5 Å resolution. Triangles, after phase extension to 3.5 Å resolution. Squares, after further phase extension to 3.0 Å resolution. (Reproduced from Rossmann et al. (1985), by permission of Macmillan Magazines Ltd.)

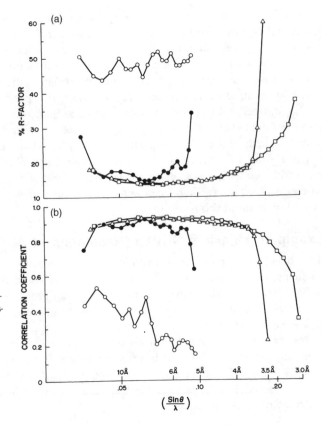

Box 10.3 Measure of performance

The performance of subunit averaging procedures may be assessed by an R factor. It assesses how well the modified electron density fits to the observed structure amplitudes. (Alternatively, the correlation between the modified and observed structure amplitudes may be calculated. This gives results that carry more statistical meaning.)

The method compares the magnitudes of structure factors for a structure on which local symmetry has been enforced ($|F_c|$) to the magnitude of the observed structure factors ($|F_o|$). To calculate F_c, the averaged electron-density distribution is placed within each subunit envelope, and the density outside the subunit envelopes may be set to zero, or to an averaged value if this is appropriate to the problem.

The standard R factor may then be calculated over defined groups of reflections h,k,l (most frequently for a series of resolution ranges).

$$R = \frac{\sum_{hkl} ||F_o(hkl)| - |F_c(hkl)||}{\sum_{hkl} |F_o(hkl)|}.$$

Alternatively the correlation between $|F_o(hkl)|$ and $|F_c(hkl)|$ may be calculated for the group of reflections. The correlation factor was defined in Box 9.1. Of course, the aim is a low R factor (near to zero) or a high correlation (near to 1).

Phase extension with high-order non-crystallographic symmetry

When there are many identical subunits, it is possible to use non-crystallographic symmetry to derive phases at gradually increasing resolution. This must be done very cautiously, to maintain a link with the structure already determined. As a rule of thumb, if the maximum diameter of the subunit is D, the reciprocal resolution $1/d_{max}$ must be increased by much less than $1/D$ at each step of phase extension.

Example: rhinovirus HRV14 (Rossmann *et al.* 1985)

In the HRV14 case, extension of resolution from 5 Å to 3 Å was carried out in 20 steps (corresponding to increasing the resolution by $1/130$ Å at each step). Each step consisted of calculating the phases corresponding to the current electron density at slightly higher resolution, and applying these phase angles to the observed structure amplitudes. These structure factors (observed structure amplitudes with calculated phases) were included in calculation of electron density at this slightly higher resolution, which was then averaged according to the non-crystallographic symmetry as before. Density outside the subunit envelopes was set to a constant value.

Figure 10.12 shows that when the phases were extended to 3.5 Å, accurate symmetry was maintained very close to the resolution limit (open triangles). When the phases had been extended to 3.0 Å (open squares), the agreement showed significant deterioration beyond about 3.5 Å. The 3.5 Å map was used for initial interpretation in terms of atomic positions, and this subsequently allowed the analysis to be further extended to 3.0 Å.

Another method: maximum entropy

The maximum entropy method of phase improvement is still under development, but has proved its power to improve electron-density maps under certain circumstances. *Maximum entropy* refers to the strategy of the method, which is to add information to a map sequentially, in such a way as to leave all possibilities open as long as possible. It is applicable where some phase angles can be determined with reasonable accuracy, but a substantial group of others remain poorly determined. This can arise, for example, if isomorphism is poor, if only one really good isomorphous derivative is available, or if the arrangement of heavy atoms possesses some misleading pseudosymmetry. The method can also start from analysis of crystals by electron microscopy.

It provides a procedure for cautiously adding phased reflections to a map calculated from a sub-set including the best-phased reflections, and using the existing map to help determine the phases of the newly added reflections.

Example: tryptophanyl-tRNA synthetase (Doublié et al. 1994)

The method was exploited successfully in structure determination of tryptophanyl-tRNA synthetase, where available heavy atom derivatives exhibited severe non-isomorphism. Crystals were grown in which selenomethionine had been substituted for all the methionines in the protein. A difference map was calculated against the parent structure, but did not reveal the selenium positions. The phases of 28 strong but poorly phased reflections were permuted to find a combination which indicated improved probability of correctness. These improved phases allowed nine selenium sites to be identified in a difference map, which finally led to the structure being determined.

'Combined' phases

It has been assumed so far that one must either use the phases directly generated from experiment (by isomorphous replacement or by anomalous scattering, for example), or use the phases generated from a structure by density modification. This is, of course, an oversimplification. A variety of procedures exist to generate phases which are a compromise between these two extremes, and it is often more effective to use these phases in further analysis.

Further reading

More advanced surveys of material described in this chapter are to be found in:

Drenth, J. (1994). *Principles of protein X-ray crystallography*, Chapter 8, pp. 183–200. Springer, New York.

Kleldgaard, M. (2000). Electron density—calculation, modification and interpretation. In *Structure and dynamics of biomolecules* (ed. E. Fanchon, E. Geissler, J.-L. Hodeau, J.-R. Regnard, and P.A. Timmins), pp. 79–101. Oxford University Press, Oxford.

Woolfson, M.M. and Fan, H.F. (1995). *Physical and non-physical methods of solving crystal structures*. Cambridge University Press, Cambridge.

Zhang, K.Y.J., Cowtan, K.D. and Main, P. (2001). Phase improvement by density modification. In *International tables for crystallography*, Vol. F. (ed. M.G. Rossmann and E. Arnold), pp. 311–24. International Union of Crystallography/Kluwer, Dordrecht.

11

Electron-density maps

When everything has been done to make the phases as good as possible, the time has come to examine the image of the structure in the form of an electron-density map. The electron-density map is the Fourier transform of the structure factors (with their phases) (Box 11.1). If the resolution and phases are good enough, the electron-density map may be interpreted in terms of atomic positions.

In practice, it may be necessary to alternate between study of the electron-density map and the procedures mentioned in Chapter 10, which may allow improvements to be made to it.

Viewing electron-density maps

Electron-density maps contain a great deal of information, which is not easy to grasp. Considerable technical effort has gone into methods of presenting the electron density to the observer in the clearest possible way.

Box 11.1 An electron-density map

The observed structure amplitudes $|F_{obs}(h)|$ may be used with the best available phases, $\alpha_{best}(h)$, using a figure of merit $m(h)$ if appropriate (see Chapter 7), to form experimentally derived structure factors:

$$F_{exp}(h) = m(h)|F_{obs}(h)| \exp[i\alpha_{best}(h)].$$

These are used in a Fourier transform calculation to create an electron density $\rho_{exp}(x)$, using Box 5.1 eqn 3:

$$\rho_{exp}(x) = \frac{1}{V}\sum_{h} F_{exp}(h) \exp[-2\pi i h \cdot x].$$

This three-dimensional function is the experimentally derived electron-density map.

The Fourier transform is calculated as a set of electron-density values at every point of a three-dimensional grid labelled with fractional coordinates x, y, z. These coordinates each go from 0 to 1 in order to cover the whole unit cell. To present the electron density as a smoothly varying function, values have to be calculated at intervals that are much smaller than the nominal resolution of the map.

Say, for example, there is a protein unit cell 50 Å on a side, at a routine resolution of 2 Å. This means that some of the waves included in the calculation of the electron density go through a complete wave cycle in 2 Å. As a rule of thumb, to represent this properly, the spacing of the points on the grid for calculation must be less than one-third of the resolution. In our example, this spacing might be 0.6 Å. To cover the whole of the 50 Å unit cell, about 80 values of x are needed; and the same number of values of y and z. The electron density therefore needs to be calculated on an array of $80 \times 80 \times 80$ points, which is over half a million values.

Although our world is three-dimensional, our retinas are two-dimensional, and we are good at looking at pictures and diagrams in two dimensions. Everyone finds it difficult to see and to remember complex three-dimensional shapes. We need every bit of help we can be given to understand them, and this inevitably requires the picture to be simplified.

In looking at an array of electron-density values, the first simplification may be achieved by contouring them. Instead of looking at the values themselves, we may look at the surfaces that separate low values of the density from higher ones.

Sometimes, it is useful to look at a two-dimensional section or slice through a three-dimensional map. It is convenient to use a contoured representation, just like that used on conventional maps of mountainous terrain (Fig. 11.1). Such a map gives full detail about the variation of density over the section.

One way to view a small piece of three-dimensional structure is to look at a series of such sections, stacked on transparent sheets (Fig. 11.2). One difficulty, which is quickly apparent, is that two of the dimensions, say x and y, have a different appearance from the stacking direction z.

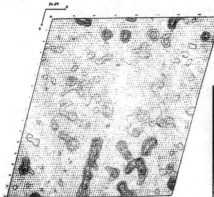

Fig. 11.1 Electron-density section hand-contoured over figure field. Normally the contours would be presented without the individual values.

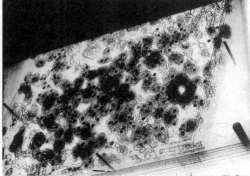

Fig. 11.2 A stack of electron-density sections, showing an α-helix and the haem of myoglobin, as obtained in 1959.

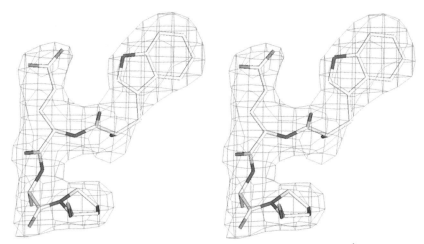

Fig. 11.3 Fragment of electron density calculated at 2.8 Å resolution from a structure determined by isomorphous replacement, showing density for Pro–Ala–Glu–Trp. This is a stereo version of Fig. 1.18. (Reproduced from Britton *et al.* (2000), with permission from Elsevier Science.)

To give a more convincing three-dimensional impression, it is possible to calculate three sets of contoured representations, on the x–y planes, the y–z planes, and on the z–x planes. The three sets of contours representing a chunk of electron density can be superimposed in three dimensions and projected on a plane, giving a clearer idea of its shape (Fig. 11.3).

There is far too much information to view the whole electron-density map in this way. Usually, only one contour level is represented at a time. (A second contour can be shown in a different colour.) A small three-dimensional block of density is presented, limited by the window or margins of the diagram in two dimensions, and by two planes which limit the third dimension as a slab. All the density outside this slab is suppressed. Views of this kind (Fig. 11.3) are frequently shown in print.

Computer displays allow the three-dimensional density to be presented more effectively, using the same three sets of superimposed contours. Slow rotation of these views by the computer quickly builds up a strong three-dimensional sensation. Over 90% of us have two good eyes, and also get three-dimensional sensation from stereoscopic effects. Systems using special glasses which cause different images to be seen by the two eyes, can present stereoscopic images. With simultaneous use of colour, rotation, and stereoscopic viewing, an impressive level of 'virtual reality' can be achieved with easily available equipment.

Stereo viewing is frequently offered in print. Two images are printed side by side, one to be viewed by the left eye and one by the right (Fig. 11.3). A holder with two lenses may be placed in front of the eyes to ensure that each sees the correct image. Many people quickly learn to view these diagrams without the aid of auxiliary lenses, so long as the two images are not too far apart. They are fairly easy to view stereoscopically if they are 50–60 mm apart, but it may be found impossible to fuse the two views if they are further apart than your eyes (normally 75 mm or so).

It is important to remember that these methods, impressively effective as they are, show a very small amount of the total information. Not only does one view a small fraction of the unit cell at any time, but also one is seeing only one contour level. Adjustment of the level used for contouring can have a dramatic effect on the appearance of the map. The electron-density level that gives the clearest representation of strong, well-ordered features is often between 1.0 and 1.5 times the root mean square density of the map. But this may fail to show up important details such as water molecules, bound substrate, or alternative conformations.

In viewing published contour diagrams, it is important to realize that the author has made many decisions about how to present it. The point of view, the volume chosen to be included, especially the depth of the slab, and the contour level all have to be adjusted carefully to give a clear representation. The author has no choice but to choose a view that clearly shows the features he wishes to present. It is usually dangerous to try to interpret other features in a published diagram. Wherever possible, further study of the electron density should use computer facilities to represent it from the original density data.

Electron density at different resolutions

The amount of detail that can be extracted from an electron-density map obviously depends on the resolution. The quality of the map also depends on the quality of the phase angles that are used to calculate it. For this reason no exact rules can be given about the resolution required, in order to define various features. Figure 11.4 shows part of a map including both protein and nucleic acid at 3.0 Å resolution.

Usually the quality of phase angles deteriorates towards higher resolution. As the mean phase error increases to over 60° (figure of merit less than 0.5), extending the resolution does little to improve the image. Thus the 'effective' resolution of a poorly phased electron-density map may be much less than its 'nominal' resolution. It is important to remember that the quality of the electron density is also destroyed by a high B factor. Parts of a map with local disorder may be uninterpretable, while other parts give detailed information.

Fig. 11.4 Part of a map at 3.0 Å resolution of a protein-nucleic acid complex, shown as a stereo pair. Base-pairs and ribose rings are easily recognized; some amino-acid side chains appear strongly, but details of the main chain conformation are hard to discern. (Reproduced from White *et al.* (1998), by permission of Nature Magazines Limited.)

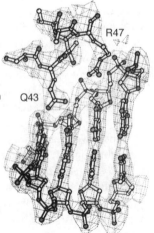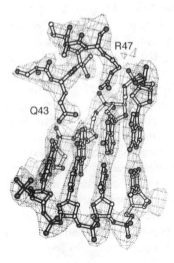

In Fig. 11.5 electron-density maps of the same structural features are presented at different resolutions. These maps are produced using structure factors calculated from atomic positions. They are 'perfect' maps at the given resolution, not degraded by experimental error. Maps generated experimentally will have poorer quality. As the resolution is enhanced, more detailed features of a well-ordered structure can be discerned. Table 11.1 gives a general indication of what can be interpreted, but should be treated with caution.

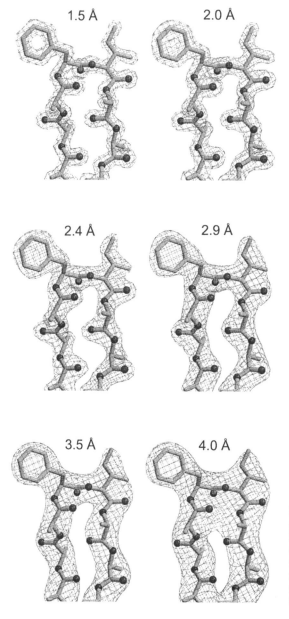

Fig. 11.5 Electron density of a β-bend presented at various resolutions: 1.5 Å, 2.0 Å, 2.4 Å, 2.9 Å, 3.5 Å, and 4.0 Å. (Diagram kindly prepared by Peter Brick, using modified Molscript (Esnouf 1997) and Raster3D (Merritt and Bacon 1997).)

Table 11.1 A rough guide to resolution required to identify features of different types in a well-phased electron density map

Type of feature	Approximate resolution
Proteins	
α-helix	9 Å
β-sheet	4 Å
'random' main chain	3.7 Å
aromatic side-chains	3.5 Å
shaped bulbs of density for small side chains	3.2 Å
interpretable conformations for side chains	2.9 Å
density for main-chain carbonyl groups, identifying plane of peptide bond	2.7 Å
ordered water molecules	2.7 Å
dimple at centre of aromatic ring	2.4 Å
correct stereochemistry at C$^\beta$ of isoleucine	2.2 Å
puckering of proline	2.0 Å
resolving individual atoms	1.5 Å
Nucleic acids	
double helix	20 Å
single strand	12 Å
stacked base pairs	4.0 Å
phosphate groups	3.5 Å
purines and pyrimidines distinguished	3.2 Å
individual bases identified	2.7 Å
puckering of ribose	2.4 Å
resolving individual atoms	1.5 Å

Example: interpretation at poor resolution (Simpson *et al.* 1998)

Figure 11.6 shows a map at 3.7 Å resolution, from a virus whose structure has been improved by averaging 30 subunits in the crystal asymmetric unit. This has evidently produced superb phasing. In this case, it was possible to trace the peptide chain for over 400 residues.

Interpretation of electron-density maps

If the resolution and phase determination are sufficiently good, map interpretation is almost obvious. Even if the interpretation is easy, it is still laborious to record the precise conformation of every amino acid, but computer assistance reduces this labour enormously.

Computer assistance is also useful when maps are hard to interpret. Computer programs can propose possible conformations of the main chain and, if these are accepted, they draw on existing structural knowledge to propose atomic positions which have been

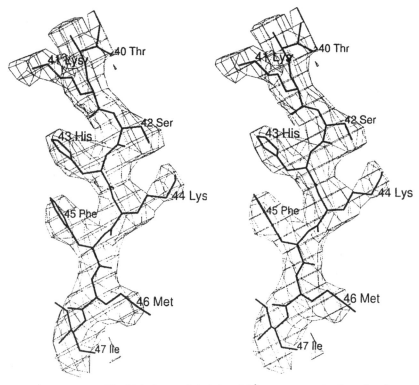

Fig. 11.6 Extended chain in a 3.7 Å map contoured at 2σ, using phases derived from 30-fold redundancy in a virus structure. (Reproduced from Simpson *et al.* (1998), with permission from Elsevier Science.)

found in similar main-chain conformations. Side chains are put into the conformations most frequently found. The results of these choices must be considered sceptically.

In the case where a sequence is available but the three-dimensional structure is being seen for the first time, it is often difficult to be sure what part of the sequence should be aligned to a particular piece of electron density. For proteins, important pointers are the positions of sulphur atoms (methionine, cysteine, and cystine), which should be the highest protein electron density at better than 3.0 Å resolution, and aromatic side chains, especially the indole of tryptophan, which should be easily recognizable if the main chain can be followed reliably. Glycines are also easy to recognize.

The direction of the main chain can be decided easily from α-helix density, because the side chains tend to project backwards to the N-terminal end of the chain. With satisfactory phases to a resolution better than 2.9 Å, the peptide carbonyl groups should be clearly visible.

Computer-derived coordinates from these 'model-building' activities are usually adjusted automatically to give bond lengths and bond angles which accurately represent conformations observed in known structures at high resolution. In a protein, assuming the amino-acid sequence to be known, the main variables defined by the model-building activity would then be the torsion angles, which represent a rotation of one part of the structure about a chemical bond. Similar considerations apply to nucleic acid structure.

Model building to define all individual atomic positions should generally be postponed until reasonably good structure factors are available to at least 3.2 Å. Interpretation at poor resolution can be greatly assisted by knowledge of a closely similar structure.

Problem regions in maps

Most proteins have local regions of more serious disorder, and during the initial intepretation of the map it is desirable to leave these uninterpreted (to avoid later problems by creating a bias towards an incorrect guess).

Errors in tracing the main chain can be particularly serious. A wrong interpretation is not easy to identify later, but may prevent further progress in structure determination. If there is doubt about the continuity of the main chain, except near the ends, further improvements to the map are needed before detailed interpretation (Chapter 10). A spectacular error is to follow a piece of chain in the reverse direction.

On the other hand, small breaks in the chain that leave no ambiguity about its continuity are not serious. These often occur where a couple of surface side chains are severely disordered. In such cases, detailed interpretation should be postponed, and the coordinates for the disordered amino acids and the linking peptide should be omitted from the interpretation for the time being. It is also common for a few amino acids at the chain termini to be completely disordered.

There are sometimes problems in the vicinity of the sites of heavy atoms used for phasing. Errors in assigning scale factors, or in the occupancy or temperature factor of a heavy atom can lead to huge features (positive or negative) in the electron density at the site. Improved computational methods for assigning these quantities have reduced this problem, but do not always eliminate it. The density for atoms close to a heavy atom site (within the distance representing the resolution or so) may be seriously distorted by these effects.

Electron-density difference maps: looking for improvements

It is always important to check the validity of a newly interpreted atomic model or model feature. This may be done by comparing an electron-density map corresponding to the built model, with the map from which it has been built. This comparison is made most effectively using an electron-density difference map, subtracting one map from another (Box 11.2).

Box 11.3 defines the most useful type of electron-density difference map for this situation. This type of map is referred to as an '$F_o - F_c$ map' in the jargon, the subscripts o and c referring to 'observed' and 'calculated' structure factors. Such a map will have positive density where features exist which have not been adequately represented in the model, and negative density where the model contains features not supported by observation. Obviously, it is essential to study negative features, as well as positive ones, in this map.

As soon as a reasonable atomic model is available for a large fraction of the structure, the phases generated by calculation from the model are likely to be better than those determined by experiment. These phases can be extended to higher resolution. In this way, improvements may be made to the model, and this can allow cyclic improvement of

Box 11.2 An electron-density difference map

Structure factors for a crystal may be calculated from a model using the structure factor formula (Box 5.2, eqn 2). These structure factors may be denoted by $F_{calc}(\boldsymbol{h})$:

$$F_{calc}(\boldsymbol{h}) = \sum_{N} f_i \exp[2\pi i \boldsymbol{h} \cdot \boldsymbol{x}_i] \exp[-B_i \sin^2\theta/\lambda^2].$$

The simplest type of difference map presents the difference between the experimentally derived electron density $\rho_{exp}(\boldsymbol{x})$ and the density $\rho_{calc}(\boldsymbol{x})$ corresponding to the model.

It is not necessary to calculate two electron-density maps ρ_{exp} and ρ_{calc} and subtract them point by point over the whole cell. A Fourier series using the structure factor difference as coefficients produces the same result. Using Box 5.1, eqn 3:

$$\rho_{exp}(\boldsymbol{x}) - \rho_{calc}(\boldsymbol{x}) = \frac{1}{V}\sum_{h} (F_{exp}(\boldsymbol{h}) - F_{calc}(\boldsymbol{h})) \exp[-2\pi i \boldsymbol{h} \cdot \boldsymbol{x}]. \tag{1}$$

The conventional $F_o - F_c$ difference map, normally used to improve map interpretation, is somewhat different from this, as discussed in Box 11.3.

Box 11.3 The conventional $F_o - F_c$ map

When an atomic model has been made, it may be considered likely that the phase angles $\alpha_{calc}(\boldsymbol{h})$ derived from it are more accurate than the experimental phases $\alpha_{best}(\boldsymbol{h})$. If this is so, it means that a more accurate electron density $\rho_o(x)$ can be calculated using:

$$F_o(\boldsymbol{h}) = |F_{obs}(\boldsymbol{h})| \exp[i\alpha_{calc}(\boldsymbol{h})]$$

as Fourier coefficients. (It is also possible to use 'combined' phases in which a weighting scheme makes a compromise between the calculated and the experimental phases.)

The map conventionally referred to as a $F_o - F_c$ map uses the calculated phase angles in this way. Instead of Box 11.2 eqn 1, the expression used to calculate this type of map is:

$$\rho_o(\boldsymbol{x}) - \rho_{calc}(\boldsymbol{x}) = \frac{1}{V}\sum_{h} (|F_{obs}(\boldsymbol{h})| - |F_{calc}(\boldsymbol{h})|) \exp[i\alpha_{calc}(\boldsymbol{h})] \exp[-2\pi i \boldsymbol{h} \cdot \boldsymbol{x}].$$

the modelling process. In initial stages, it is not usual to try to assign individual B factors to different parts of the model. As well as showing wrongly placed atoms, the difference map will show negative density where parts of the structure are more doubtful due to disorder. The first calculated difference map may be expected to have many features.

Further rounds of model building

Chapter 12 explains how an atomic model can be fitted to the data in a more quantitative way, known as structural refinement. Refinement can only begin if the starting model has most of its atoms close to their correct position. Even then, badly placed atoms may not be brought back to their proper positions, and it is better to leave them out at first. Therefore further rounds of model building are usually necessary after refinement, and it is always important to use difference maps to check whether they are needed.

By now the phase angles calculated from the model are almost certainly superior to those derived experimentally, and a map using them might be more accurate. However, the use of phases calculated from the model inevitably creates electron density which is biased towards confirming the model. In order to minimize this bias, the effect is reduced (but never eliminated) by calculating the maps explained in Box 11.4. Usually a '$2F_o-F_c$ map' is calculated, but occasionally variants such as the '$3F_o-2F_c$ map' are employed. These maps look like ordinary electron-density maps of the protein, but considerably reduce the bias introduced from the model. When publishing an electron-density map, to demonstrate the quality of the electron density, a $2F_o-F_c$ map has normally been used. Recently model bias has been further reduced by a technique known as sigma-weighting (see Further reading).

Box 11.4 Reducing model bias in electron-density maps

Phase angles are of overwhelming importance in the calculation of electron density. The phases from the model density tend to reproduce the model structure. An electron-density map in which the structure amplitudes $|F|$ are all set to 1, but the phases are retained, looks remarkably like the starting structure.

At an early stage of map interpretation in terms of atoms, the phase angles, α_{calc}, become more accurate than the experimentally determined phase angles. But electron-density maps using model phase angles are automatically biased to re-create the model density.

To remove some of the bias from the calculated electron density, the $2F_o-F_c$ map is often used. Since

$$(2|F_{obs}(\boldsymbol{h})| - |F_{calc}(\boldsymbol{h})|)\exp[i\alpha_{calc}(\boldsymbol{h})]$$
$$= |F_{obs}(\boldsymbol{h})|\exp[i\alpha_{calc}(\boldsymbol{h})] + (|F_{obs}(\boldsymbol{h})| - |F_{calc}(\boldsymbol{h})|)\exp[i\alpha_{calc}(\boldsymbol{h})]$$

this can be recognized as an electron-density map using calculated phases (first term) with an added electron-density difference map (second term). The difference map negates unsupported features of the calculated map and strengthens the F_o map, thus reducing bias.

Some workers prefer to remove the bias more strongly and use other Fourier coefficients, such as $3F_o - 2F_c$. If one goes too far in this direction (such as $11F_o - 10F_c$), the map approximates to an electron-density difference map.

Features observed in maps, but not modelled

Electron-density maps often reveal features that are not part of the protein. These features include ordered molecules from the solvent and ligands such as enzyme substrates. These are presented most clearly in an $F_o - F_c$ map.

A $2F_o - F_c$ map is also useful in generating enhanced density in parts of the structure that have not so far been interpreted, since the F_c map will have no significant density in this part. The $2F_o - F_c$ density will have a double contribution from the observed data. Uninterpreted features therefore emerge at enhanced level in a $2F_o - F_c$ map, or even more strongly in a $3F_o - 2F_c$ map. An example is shown in Fig. 11.7.

It is usually prudent to omit solvent molecules from the structure factor calculation until the macromolecule is fully interpreted. Otherwise an incorrectly assigned solvent molecule, derived from background error in the map, may be difficult to remove.

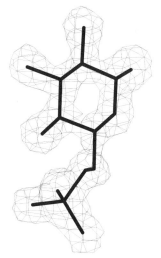

Fig. 11.7 Density for an enzyme reaction product (glucosamine-6-phosphate) shown at 1.57 Å resolution in a $3F_o - 2F_c$ map which did not include it. (Reproduced from Teplyakov *et al.* (1998), with permission from Elsevier Science.)

Omit maps

If interpretation of some parts of the map is still doubtful, more thorough elimination of model bias may be achieved by calculation of an 'omit map'. In this case, a volume of the map is defined, which requires more thorough investigation. All the model atoms that lie in this volume, or close to it, are deleted from the model. The phase angles calculated from the remainder of the structure are certainly less accurate, but are considered to give an unbiased estimate of phase angles for calculation of the density in the volume omitted.

In desperate cases, where the whole model seems unreliable, a complete new electron-density map can be constructed from omit maps. It is essential that, in calculating each part of the density, only a small fraction of the model volume is omitted (one-eighth at most, preferably less). A moment's thought will show that even such a map is undoubtedly derived from phases calculated from a model, and even these procedures can hardly be unbiased.

Example: experimental map compared with final $2F_o - F_c$ (Yu *et al.* 1998)

The experimental electron density obtained from a MAD map at a nominal resolution of 2.4 Å is shown in Fig. 11.8a. In the 2.4–2.5 Å shell, the MAD phasing power was below 0.5 for all wavelength comparisons, and the figure of merit was 0.25. Therefore, although reflections were included to 2.4 Å, the poor phasing in the higher-resolution part of the data means that the effective resolution is considerably worse. The density shown includes some protein and an ATP molecule, and the final interpretation is superimposed on the map. It appears as though the detailed conformation of many side chains would be uncertain from this map, and peptide carbonyl density would be difficult to recognize.

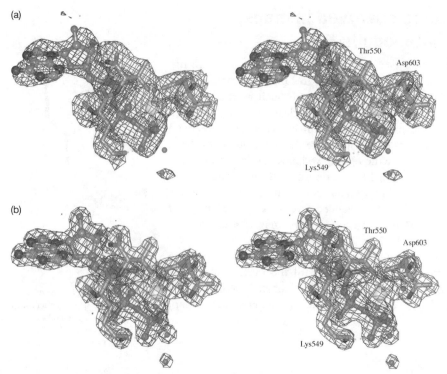

Fig. 11.8 Oligomerization domain of *N*-methylmaleimide-sensitive factor shows (a) the experimental MAD map at nominal 2.4 Å resolution, and (b) $2F_o - F_c$ map using refined phases to 1.9 Å, both contoured at 1.25σ (Reproduced from Yu *et al.* (1998), by permission of Nature Magazines Limited.)

The authors used solvent flattening, histogram matching, and phase extension to help with the interpretation. After structural refinement, phases were calculated from an atomic model to 1.9 Å resolution, the limit of intensity measurement. This generated a greatly improved $2F_o - F_c$ map, seen in Fig. 11.8b. All atomic positions are covered by density, side chains are much more clearly resolved, a ribose oxygen is neatly fitted, and ordered water molecules are clearly identified.

Example: use of an omit map (Daniels *et al.* 1998)

In Chapter 10 the effects of solvent flattening, histogram matching, and 2-fold symmetry averaging on an experimental map were demonstrated by an example of a PDZ domain from hCASK.

After improvement and refinement, this structure gave a map at 2.7 Å resolution, part of which is shown in Fig. 11.9a. To confirm that difficult parts of the structure had been correctly interpreted, an omit map was calculated. This map used combined phases (see Box 11.3), merging the experimentally derived phase information from MAD with phases from a partly refined model. The omit map (Fig. 11.9b) was calculated at 2.1 Å, and contoured at a different level, so it is not strictly comparable with Fig. 11.9a, but it shows

(a)

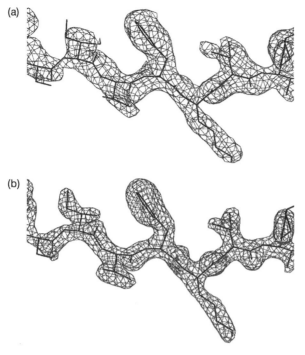

(b)

Fig. 11.9 Electron density for a few residues of the PDZ domain of hCASK. (a) Starting from MAD phases, after solvent flattening, histogram matching, and 2-fold symmetry averaging at 2.7 Å resolution (as Fig. 10.3b). (b) Omit map, using phases from the MAD observations combined with the phase information from a partially refined model, and extended to 2.1 Å from the model. (Reproduced from Daniels *et al.* (1998), by permission of Nature Magazines Limited.)

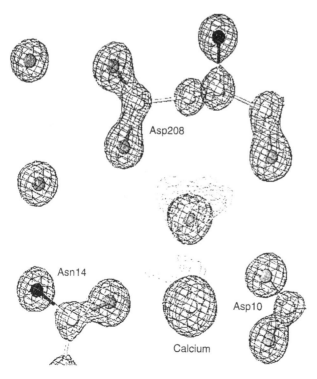

Fig. 11.10 Part of the density at 0.94 Å resolution for concanavalin A, showing Asn14 and nearby ordered water molecules. The darker contours show the electron density as a $(2F_o - F_c)$ map, in which each protein atom is clearly resolved. The lighter contours show a $(F_o - F_c)$ difference map. Hydrogen atoms were not included in the model and the difference map clearly reveals hydrogen positions. (Reproduced from Deacon *et al.* (1997), by permission of the Royal Society of Chemistry.)

considerably improved quality. Density for the main-chain carbonyl oxygen atoms is clearly seen.

Example: very high resolution (Deacon *et al.* 1997)

At very high resolution the electron densities of atoms are fully resolved from each other, and electron density belonging to hydrogen atoms begins to become discernible (Fig. 11.10).

Further reading

Little can be found on this topic in textbooks since the comprehensive discussion in:
Blundell, T.L. and Johnson, L.N. (1976). *Protein crystallography*, Chapter 13. Academic Press, New York.

Meanwhile the severe practical problems have been tremendously alleviated by computer procedures, which can be performed so easily that the prime movers in simplifying them now have to emphasize the possibility of mistakes. See:
Brandén, C.-I. and Jones, T.A. (1990). Between objectivity and subjectivity. *Nature*, **343**, 687–9.

Kleywegt, G.J. and Jones, T.A. (1995). Where freedom is given, liberties are taken. *Structure*, **3**, 535–40.

Reviews of several aspects appear in:
Carter, C.W. and Sweet, R.M. (ed.) (1997). Macromolecular crystallography (Part B). *Methods in Enzymol.*, **277**.

especially:
Jones, T.A. and Kjeldgaard, M. (1997). Electron-density map interpretation. *Methods in Enzymol.*, **277**, 173–207.

Kleywegt, G.J. and Jones, T.A. (1997). Model building and refinement practice. *Methods in Enzymol.*, **277**, 208–29.

Kleywegt, G.J. and Jones, T.A. (1997). Detecting folding motifs and similarities in protein structures. *Methods in Enzymol.*, **277**, 525–45.

The sigma-weighting procedure is described in:
Read, R.J. (1987). Model phases: probablilities and bias. *Methods in Enzymol.*, **277**, 110–28.

12

Structural refinement

At this stage, we have derived a model from an electron-density map and have interpreted it as closely as we can in terms of molecular structure. Provided the job has been done well enough, the next task of improving that interpretation can be left to computational procedures known as *structural refinement*. If further uninterpreted features of the structure are revealed, it will be necessary to go back to the methods of Chapter 11 to improve the interpretation.

The purpose of structural refinement is to adjust a structure to give the best possible fit to the crystallographic observations. The intensities of the Bragg reflections constitute the observations, and the various quantities that define the structure are adjusted to give the best fit. Box 12.1 gives an outline of what is meant by refinement of quantities to fit observations. In structural refinement, a measure of the discrepancies between the calculated X-ray scattering by the model structure and the observed intensities is defined: this is called the *refinement parameter*. The purpose of the refinement procedure is to alter the model to give the lowest possible refinement parameter.

Box 12.1 uses a simple example to bring out some important general points:

1. My model will be specified by a number of variables. In a diffraction experiment, they are usually the coordinates and B factor of every atom. If the number of observed quantities is less than the number of variables, the results can have no validity.
2. If the number of observations equals the number of variables, a perfect fit can be obtained, irrespective of the accuracy of the observations or of the model. (This is true of so-called linear problems, and approximately so in non-linear cases.)
3. If the number of observations exceeds the number of variables by only a small quantity, the estimate of the reliability of the model is questionable.

In practice, refinement procedures can only work when there is a sufficient number of observations which are sufficiently accurate. Also, the model must already be good enough to make the refinement procedure meaningful. If atomic positions are to be refined, most of the atoms of the model structure must already lie within the cloud of density that represents their position.

Box 12.1 Refining parameters to fit observations

As a simple example, suppose my results require two parameters to represent them. This is the problem of fitting a straight line

$$y = mx + c$$

to a set of measurements that observe the two variables x and y. I can plot them on graph paper as (x_1, y_1), (x_2, y_2), etc. The two parameters are the slope, m, of the line and the intercept, c, which defines where the line lies on the graph. The idea that the results fit on such a straight line is the *model* which the observations can test.

If I only have one observation, it is obviously possible to put a straight line through it, running in any direction. It is absurd to try to determine m and c, and there is no information whether the model is correct.

If I have two different observations, I can always put a unique straight line through them. Thus if (x_1, y_1) is (2,3) and (x_2, y_2) is (3,4), then $m = 1$ and $c = 1$ fits it exactly (Fig. 12.1). The results look perfect, but they give me no assurance that my measurements or my model are accurate. I could get a perfect straight line by using random numbers for x and y.

It is only when I have made three observations that I begin to know whether my model is valid, that the observations all lie on a straight line. If, for example, I happen to measure (x_3, y_3) as (4,5), all three points are exactly on a straight line and I can begin to think my model is pretty good (Fig. 12.2). If I measure (x_3, y_3) as (4,6) (Fig. 12.3), I can judge a reasonable line which goes between the points. Ideally this line is the same distance from all three points, and this difference gives me an idea how well the model fits the observations. I test the model by considering whether this is the level of accuracy to be expected from my techniques. Note that although I have made three pairs of observations, this only provides a single quantity to guess their accuracy. This guess could be made much more reliable by making further observations.

It is usual to assess the accuracy of observations (and of a model to explain them) by calculating the square of the difference between the observed quantities and the values that the theoretical model suggests. The *method of least squares* finds the model parameters that minimize the sum of the squares of the discrepancies between the model and the different observations. This method is generally used in X-ray crystallographic refinements.

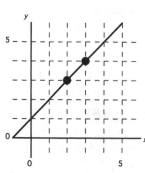

Fig. 12.1 Perfect fit to two points.

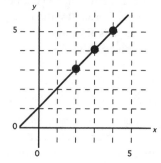

Fig. 12.2 Perfect data can give a perfect model.

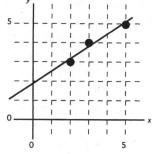

Fig. 12.3 Real data show how well the model fits.

General principle of refinement: a flea in a bowl

Refinement methods search for a minimum value of a quantity which depends on several variables. Suppose, for example, I am a flea trying to hop my way to the bottom of a bowl. The bowl has a two-dimensional surface, and the height where I am standing depends on two coordinates. From my current position on its surface, it would be a good idea to go downhill. I could estimate how far to jump depending on the slope and curvature. If the bowl is exactly circular (Fig. 12.4), downhill is definitely the correct direction, but even if it is not, I can look at the slope and curvature where I land next, and make a new decision for my next jump. If I jump too far, the slope will be in an opposed direction, and I will have to jump back.

If the surface has a narrow, twisting depression, like a river leading to a lake (Fig. 12.5), I will easily find my way near to the groove in the bottom, but it will be more difficult to find where to go along the groove.

Suppose I was not on a bowl, but on a tart-baking tray with a dozen depressions. I might find one lowest point, but I would know nothing about the eleven other possibilities— one might be slightly deeper. If I was in a shallow depression with a textured surface, say a pebbled hollow, I would often jump in wrong directions, leading me into the groove between two pebbles. By jumping always downhill, I ultimately find a lowest point, but it may be an insignificant hole in a crevice between a few pebbles.

So far the height has depended on two variables (say x and y on a piece of graph paper under the bowl). The same methods can be used in a case where the 'height' (the refinement parameter) depends on dozens or thousands of different variables. Mathematicians have devoted great effort to finding efficient ways to solve this problem, and to simplify the calculations.

Fig. 12.4 Flea jumps in a bowl.

In macromolecular crystallography we have many thousands of observed intensities. They must be used to find the best fit to thousands of atomic positions. Each of the variables defining the structure is like another dimension to the space which must be explored in seeking the best fit.

To summarize—the processes have great mathematical complexity. A strategy must be adopted which does not look at irrelevant detail until the coarser details are resolved. In the end I shall find a local minimum, but there may be much deeper ones that I shall never know about. I can try to avoid overlooking deep minima by caution in enhancing the level of detail.

First quality indicator for the structure: the *R* factor

In previous chapters, a number of reliability factors, or R factors, have been introduced as measures of the quality of the intensity measurement, isomorphous replacement,

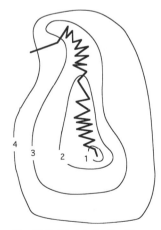

Fig. 12.5 A twisted valley. Many steps down the line of steepest slope may be needed to find the minimum.

Box 12.2 The R factor

The 'conventional' R factor compares the observed structure amplitudes $|F_{obs}|$ to those calculated from the current model $|F_{calc}|$. It is defined as

$$R = \frac{\sum_h \left| |F_{obs}| - |F_{calc}| \right|}{\sum_h |F_{obs}|}. \tag{1}$$

As with other R factors, some authors express it as a percentage. Thus '$R = 20\%$' is the same as '$R = 0.2$'.

The R factor is calculated over a group of reflections h, which may be all the observed reflections, or a particular group. Frequently an R factor is calculated over small ranges or 'bins' of resolution, to give an idea of the performance of the model as resolution is increased.

molecular replacement, and so on. The original R factor (proposed in 1945 by Alan Booth), which is implied when there is no subscript to the R, is used as a measure of the fit of the calculated diffraction by the structure to the observed intensity data. This R factor is the ratio

$$\frac{\text{sum of differences between observed and calculated structure amplitudes}}{\text{sum of observed structure amplitudes}}$$

(Box 12.2).

The R factor falls towards zero as the observed and calculated structure amplitudes agree more closely, but the values achieved depend on the resolution and degree of order of the crystals, the quality of the diffraction data, and so on.

In early stages, the R factor may be as high as 0.5. At this stage, all atoms are usually included with full weight, the coordinates are not optimized, and individual B factors have not been assigned. Even if the coordinates were perfect, the failure to assign proper B factors would have a serious effect on the R factor. The progress of refinement may be watched by monitoring the R factor, and ensuring that it continues to fall.

Classical structural refinement

In structure determination for small molecules, the refinement parameter is the agreement of the observed structure amplitudes with the scattering calculated from the atomic positions (and the associated B factors).

Before the methods explained in the next section were introduced, it appeared as though structural refinement of macromolecules could only be possible if the number of observed reflections was considerably more than the number of coordinates which define the structure. This required a resolution better than was available for the vast majority of structures being studied at that time.

Example: rubredoxin (Watenpaugh *et al.* 1973)

The 6 kDa metalloprotein rubredoxin, which has well-ordered crystals and a comparatively small unit cell, was the first protein to be refined accurately by classical refinement methods. Only reflections of intensity exceeding 2 standard deviations were included, which gave 5005 intensity measurements to 1.5 Å resolution.

Interpretation of a density map at 2 Å resolution from MIRAS phases allowed 401 of the 424 protein atoms to be placed in density, together with 23 water molecules, giving $R=0.372$. The first stage of the structural refinement used three-dimensional $(F_o - F_c)$ difference maps. Each atom was moved, just like the flea, down the steepest gradient of the difference map. It was obvious that some atoms were better ordered than others, and they were assigned different B factors over four cycles of this process. At each stage, a few atoms were added where they showed up more clearly and, if necessary, some were removed where interpretation was less certain. Progress of the refinement is summarized in Fig. 12.6.

Further refinement used the method of least squares, but since operations on 2000 × 2000 arrays were impracticable at that time, serious approximations had to be made. In each round of refinement, about a dozen cycles of least squares was followed by rebuilding the model from a difference map. Three rounds of refinement reduced R to 0.132, with 2232 variable parameters.

This produced the most accurate model for a protein molecule at that time, but close examination showed that some bond lengths were unreasonable. Clearly, some of the reduction in the R factor had been achieved by moving atoms to slightly incorrect positions. The ratio of measurements to refinable variables was 5005/2232 or 2.2. This factor will be called the *safety factor*.

Ways to describe macromolecular structure

In classical structure determination, there were four variables—x, y, z, and B—to be determined for each atom. (There could be more, for anisotropic disorder and other detail.) As we shall see, the limitations of protein crystals mean that it is hardly ever

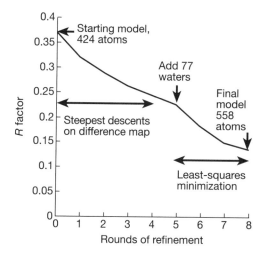

Fig. 12.6 Progress of rubredoxin refinement (data from Watenpaugh *et al.* 1973).

desirable to begin refinement at such a detailed level. Fortunately, there are ways of describing a model structure using fewer variables than four per atom.

The crystal structures of individual amino acids and small peptides have been determined with great accuracy, and the bond lengths and bond angles of the amino-acid side chains in proteins vary little from these values. Therefore, the refinement can be forced to keep to them.

In the same way, the geometry of the peptide linkage is well determined and is composed of atoms in a plane (Fig. 12.7). This geometry may be enforced in the model. The atoms that form part of an aromatic system can be forced to lie in a plane, as can other planar groups, such as the guanidinium group.

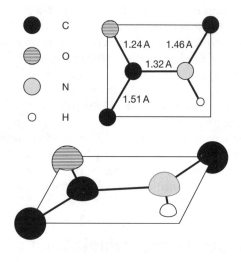

Fig. 12.7 Atoms of the peptide linkage may be made to lie in the same plane.

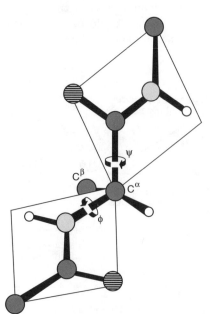

Fig. 12.8 Two consecutive peptide bonds linked at the α-carbon atom. The relative positions of all the atoms shown are specified by the Ramachandran angles, φ and ψ.

Building a model under these constraints of bond lengths, bond angles, and planar groups means that the complete structure may be specified by a series of conformational angles (sometimes called *torsion angles* or *dihedral angles*). Each angle specifies the orientation of a group of atoms relative to the structure previously specified.

Because the atoms of the peptide linkage lie in a plane, the conformation of a protein main chain can be specified by two conformational angles, ϕ and ψ at each α-carbon atom. This description was introduced by G. N. Ramachandran (Ramachandran *et al.* 1963), and these angles are generally referred to as *Ramachandran angles* (Fig. 12.8). Assuming fixed bond lengths, bond angles, and planar groups, these two angles fix the positions of all the main-chain atoms, plus the $C\beta$ atom. The conformation of the rest of the side chain can be specified by further conformational angles χ_1, χ_2, and so on, depending on the length of the side chain.

The number of dihedral angles to define a side-chain conformation varies from 0 (for glycine and alanine) to 5 (arginine), with an average of about 2. The whole structure may be defined by about 4 variables per amino acid, two for the main chain and two, on average, for the side chain.

The conformation of a nucleotide may be specified similarly, but in this case the backbone conformation needs six angles to specify it (Olson and Flory 1972; Saenger 1984).

These examples show how the structure of a complex biopolymer may be specified in great detail, using much fewer values of conformational variables, because bond lengths and angles may be considered as known quantities.

Structural refinement with restraints

In refining a macromolecular structure at moderate resolution, there are not enough observations to allow refinement of all the atomic positions independently. Instead, we might define the whole structure in terms of its dihedral angles, and use these as the variables in refinement. This, however, creates a problem in adjusting the model.

If I try to adjust a model made of components with absolutely rigid lengths and angles, by twiddling the dihedral angles about the bonds, I will often find its conformation locked. I can only make the adjustment by slightly distorting the bond angles. If the components are rigid and I apply too much force, I will simply break the model.

Exactly this problem arises in the mathematical procedures for refining a structure. It could mean that the procedure fails to make a simple adjustment which is obviously needed. So, instead of making the model absolutely rigid, it must be given a certain flexibility.

In refinement this is achieved by allowing bond angles and bond lengths to change, but by imposing a large penalty if they are changed very much. In the mathematical procedure, the refinement parameter has a component which is significantly increased for a small distortion from ideal bond lengths or angles. The bond lengths and angles are said to be *restrained*. Other restraints may also be applied to achieve a physically reasonable model, such as enforcing the planarity of groups such as peptide linkages and aromatic rings.

The introduction of restraint penalties allows quite sophisticated rules to be made. One very important rule is that atoms must not overlap. If atoms are brought too close to each other, they repel each other strongly. A system of restraints can be created which simulates the repulsion between atoms that are too close. Many other restraints can be applied to make the refinement procedure reflect real constraints on structure.

Box 12.3 indicates how these restraints are applied in the refinement procedure, and explains that they have the apparent effect of reducing the number of variables that can be refined. There is always a danger of arguing in a circle. If I restrain a structure in a certain way, because that is how I know it must be, it can be no surprise that the results are in agreement with the restraints I have applied.

Box 12.3 Restraints

Refinement attempts to improve the agreement of available data with the model, by adjusting the variables that define the model. In X-ray diffraction the most obvious observable quantities are the observed X-ray scattering, which must be made to agree with the scattering calculated from the structure, that is,

$$|F_{obs}| = |F_{calc}|$$
$$= |\sum_N \{f_i \exp[2\pi i(hx_i + ky_i + lz_i)] \exp[-4B_i \sin^2\theta/\lambda^2]\}|.$$

(see eqn 1, Box 5.2). This is called an *observational equation*. It includes variables x_i, y_i, z_i, B_i for every atom in the structure. There is an observational equation of this form for every reflection in the refinement.

To enforce it, the refinement parameter includes a term

$$\sum_h w_h(|F_{obs}| - |F_{calc}|)^2, \tag{1}$$

where w_h is a weighting term, which can be adjusted according to the accuracy of the observation.

Restraints are added by including other available data in the form of further observational equations. For example, since we know that bonds between atoms i and j always have a length l_{ij}, the observational equation would be:

$$l_{ij} = [(x_i-x_j)^2 + (y_i-y_j)^2 + (z_i-z_j)^2]^{1/2}.$$

Such an equation may be set up for every pair of bonded atoms. The refinement parameter includes a term:

$$\sum_{\text{bonded atoms}} w_{\text{bond length}} \{l_{ij} - [(x_i-x_j)^2+(y_i-y_j)^2+(z_i-z_j)^2]^{1/2}\}^2.$$

Other restraints may be applied in a similar fashion.

In this scheme, there are four variables x_i, y_i, z_i, B_i for all the N atoms in the structure. But in the mathematical formulation, restraints increase the number of 'observational' equations that must be satisfied in minimizing the refinement parameter. Applying a restraint is like increasing the number of observations, and in this way the number of observational equations can be increased so that there are many more equations than the $4N$ variables.

The bond length restraints will be applied tightly if the weighting factor $w_{\text{bond length}}$ is sufficiently large. In practice, the refinement can produce results where the bond lengths are quite rigorously enforced. In the main text we shall pretend that certain restraints are so nearly rigid that the number of refinable variables is reduced.

Box 12.4 suggests various schemes for restraining protein models. If few variables must be used, the structure might be represented just by its main-chain conformation, specified by ϕ and ψ, with side chains either omitted or inserted in a standard conformation. At intermediate resolution, the scheme of restraining bond lengths and angles, and refining with the dihedral angles as variables, may be used. At high resolution, the structure of a protein could be refined like rubredoxin, with three atomic coordinates for each atom and a separate B factor for each atom, and with no restraints on bond lengths and angles. Hydrogen atoms (which scatter X-rays very weakly) are usually considered to be in positions fixed by other atoms, but at the highest resolution they may also be included in refinement. Refinements at resolution better than 1 Å, with all restraints relaxed, for a few supremely well-ordered examples, now suggest tiny improvements to the peptide geometry which was originally defined from the study of small peptides.

Box 12.4 Numbers of refinement variables

At the lowest resolution one might try to refine only the main-chain conformational angles. There would then be just two variables, ϕ and ψ, for each amino acid.

At slightly better resolution, one might refine side-chain conformational angles χ, but tightly restrain all bond lengths and bond angles, maintain the planarity of all unsaturated systems, and have one overall B factor for the whole structure. This implies on average about four variables per amino acid.

Further improvement of resolution could allow relaxation of the bond angle at the α-carbon atom (which is known to be more flexible), and two separate B factors for each amino acid (one for main-chain atoms and the other for side-chain atoms), giving three more variables per amino acid.

At higher resolution, bond angles and ultimately bond lengths can become unrestrained, giving about 2–4 variables per non-hydrogen atom.

At the highest resolution, atoms may be allowed anisotropic B factors. These allow the density distribution of atoms to show the particular direction in which they are more free to move. Each atom's shape is described by six B factors, more than doubling the number of variables.

With four variables per amino acid, the structure of a smallish protein of 20 kDa with 180 amino acids is defined by about 720 variables. Refinement of the conformational angles is meaningless unless more than 720 intensities have been measured. Using the 3-fold safety factor suggested in the text, a minimum of 2160 intensity observations is desirable.

Taking an average Matthews volume, the minimum resolution for a refinement of this kind can be calculated from the equation in Box 5.6. It comes out at

$$d_{min} = \left[\frac{2\pi}{3} \times \frac{(\text{Matthews volume}) \, (\text{molecular weight})}{(\text{safety factor}) \, (\text{number of refined variables})} \right]^{1/3}.$$

This formula is applied to suggest appropriate resolutions for various refinement strategies in Table 12.1, using the recommended safety factor of 3.

Strategy for structure refinement: getting started

As explained in an earlier section of this chapter, according to the precision of the starting model, the strategy should be to begin with a relatively coarse-grained refinement at relatively low resolution. This will avoid the risk of becoming stuck in an insignificant depression in the terrain, far from the global minimum of the refinement parameter that I am seeking. The refinement estimates depends on phases calculated from a model, but should begin near the resolution to which phases have been determined experimentally.

This may sound simple, but is not always easy, especially for structures where, for whatever reason, experimental phases are limited to resolution worse than 3 Å. Table 12.1 gives a rough indication of the resolution at which different schemes of refinement are feasible. The indications are only rough because there is no hard-and-fast rule about the excess of observations required over variables. A resolution can be defined at which the number of variables equals the number of observations. This is the mathematical minimum requirement for refinement of these variables, but in practice a larger number of observations is essential. A suggested safety factor is to require three times as many observations as variables (assumed in the third column of the table). This safety factor is arbitrary and depends on observational accuracy and the quality of phase determination used for the starting model. It is obviously affected by the accuracy of the intensity measurements and phases, which may become very poor before reaching the nominal resolution.

At resolution worse than 3.2–3.4 Å, even with good phases, it may be difficult to follow a main chain correctly in the electron density. So a starting main-chain model may not be accessible. (An exception arises in the case of α-helices, recognizable at much poorer resolution. Haemoglobin and myoglobin were exceedingly fortunate choices for early X-ray diffraction studies.)

Table 12.1 Estimate of resolution required for different types of refinement

Variables refined	Estimated number of variables	Number of reflections[a]	Corresponding resolution[b] (Å)
Main chain ϕ and ψ only	360	1080	4.5
ϕ, ψ and side-chain χ	670	2010	3.7
ϕ, ψ, χ, angle at C$^\alpha$, two B values per amino acid	1210	3630	3.0
ϕ, ψ, χ, angle at C$^\alpha$, separate B values for each atom	2250	6750	2.5
All bond angles released, separate B values for each atom, some water included[c]	4000	12 000	2.0
Bond lengths and angles released, separate Bs, some water[c]	6160	18 480	1.76
Anisotropic B factors added	13 860	41 580	1.34

This table is only a crude guide. The estimates assume a small protein of 20 kDa, which would have about 180 amino acids and about 1400 non-hydrogen atoms, but the results are typical for other proteins. The last column depends on the Matthews volume, which is assumed to be 2.4 Å^3 per dalton.

[a] Applying arbitrary safety factor of three observations for each refined variable.

[b] Using eqn 1, Box 5.3, assuming complete data and $V_M = 2.4$ Å^3.

[c] Assuming 1 water molecule for 10 protein atoms.

Comment on rubredoxin (Watenpaugh *et al.* 1973)

Table 12.1 suggests that refinement of four values per atom should be 'safe' at 1.76 Å. We saw above that the safety factor for rubredoxin was only 2.2 at 1.5 Å resolution. The discrepancy arises chiefly because only about 70% of the reflections to 1.5 Å were included in the refinement. The rest were excluded by the cautious policy that intensities should exceed 2 standard deviations. Modern practice is to include all reflections, but this discrepancy underlines that the crude estimates in Table 12.1 assume good-quality data to the resolution limit. Including reflections in an outer shell where measurements are poor does not do much to enhance the resolution.

Once a map can be interpreted so that most of the atoms are placed in density, refinement can begin. A first refinement process may be chosen according to the guidelines of Table 12.1.

Example of conventional structure refinement: *Erythrina* trypsin inhibitor (Onesti *et al.* 1991)

In this example a trypsin inhibitor of 171 amino acids was refined, starting from 2.7 Å resolution. Refinements of this nature are now so routine that details are rarely given in publications.

The initial MIRAS map at 2.7 Å resolution included five discontinuities in the polypeptide chain. In the initial interpretation of this map, 33 amino acids were totally omitted, and 18 more were included as alanine because of the poor quality of side-chain density. The first two rounds of refinement used the 2.7 Å data originally phased by isomorphous replacement. Bond lengths and angles were restrained throughout the refinement. Difference maps were examined and manual adjustments of the model were made after each 'round' of refinement.

From round 3 all the recorded intensities to 2.5 Å were used and more amino-acid side chains were gradually included. Five C-terminal amino acids appeared too disordered to be modelled. By the end of round 8, the other 166 amino acids and 60 ordered water molecules had been included in the model, and the overall R had dropped to 0.208.

In this refinement bond lengths and angles were tightly restrained, so that the structure was described by about 2200 variables. The refinement used 7770 reflections, giving a

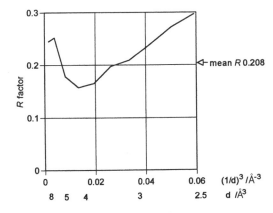

Fig. 12.9 Variation of R with resolution after refinement of *Erythrina* trypsin inhibitor. The horizontal axis includes equal numbers of reflections in equal widths. (Data from Onesti *et al.* 1991.)

safety factor of 3.6. Figure 12.9 shows how the final R factor varies with resolution, falling below 0.16 at around 4.0 Å resolution, but rising to 0.30 at the resolution limit of 2.5 Å.

Comment

The starting model was conservative and omitted any conjectural details. Consequently, the refinement seems to have avoided being driven into a false minimum.

Rigid body refinement

It may be difficult to start conventional refinement from a model based on map interpretation. Another method may be available if the model is believed to include blocks of predictable structure, such as domains similar to known structures. It is possible to refine lengths of secondary structure, α-helices, β-strands, or helical oligonucleotides, in the same way. In the model, these units are placed as well as possible into the electron density. This may require the use of molecular replacement techniques described in Chapter 9. By treating each unit as a rigid body, refinement may move it to give the best fit to observed intensities.

This refinement is mathematically just like the refinement presented in Box 12.2. But if a unit is rigid, all the distances between its atoms are fixed, whether or not they are directly bonded to each other. There are only about six variables for each block of structure.

Rigid body refinement of this kind has often been used in difficult cases, to get the refinement started, as in the example of metabotropic glutamate receptor at the end of the chapter.

Carry on refining

When refinement approaches convergence (negligible further reduction of the refinement parameter), the resolution can be increased. But before extension of resolution it is very important to look closely at a difference map (its negative features as well as the positive ones!), which may show obvious errors that the refinement procedure has failed to correct. If this is not done, the refinement is in danger of being stuck in a false minimum.

Once the positions of nearly all atoms have been assigned confidently, the resolution can be extended cautiously to the highest available.

If the refinement does not appear to be advancing properly, study of a difference map is the safest way to proceed. An alternative method is equivalent to shaking up the structure, and allowing it to settle down in a better way. The jumping flea finds the surface he has settled on uncomfortably hot, and jumps high in the air. He will land some distance from the local minimum he found. As the surface temperature is gradually cooled, he makes smaller jumps, always downhill. This procedure may lead him to a lower minimum. Computationally, all the atoms are given a small random displacement to positions where the refinement parameter is inevitably higher, and then refined from there. This method is known as *simulated annealing*.

In the above example of the *Erythrina* trypsin inhibitor refinement (Onesti *et al.* 1991) a step of simulated annealing was tried after the first round of refinement and reduced R from 0.321 to 0.297. This step fortunately 'flipped' one peptide bond by 180°. In this case, a later trial of simulated annealing gave negligible improvement.

The free *R* factor

Refinements have often been monitored by following the fall of the *R* factor, but this criterion was criticized by Axel Brünger, who demonstrated the bias that is introduced by using the same data for monitoring as for refinement. The refinement procedure may make spurious adjustments which reduce the value of *R*, but which do not actually improve the model.

It is now usual to select randomly a small fraction of reflections (typically 5%) which are deleted from the data used for refinement. These reflections, and only these, are used to calculate an *R* factor, which is identified as R_{free} (Box 12.5). Reduction of the R_{free} factor is an unbiased estimate of the improvement of the model, because R_{free} is calculated from reflections that the refinement procedure does not 'know' about. For this reason R_{free} is somewhat larger than the conventional *R* factor. As always, cases vary, but typically R_{free} is about 1.2 times *R* at the end of refinement.

The *R* factor varies in any case with resolution. It is often relatively high at low resolution (say as far as 8 Å). This indicates that the methods used for modelling the solvent part of the density are imperfect. This can be corrected, but rarely has a serious effect on interpretation at the level of individual amino acids. *R* reduces to its lowest value at intermediate resolution, and begins to rise for the weaker reflections at higher resolution (Fig. 12.9). If it rises beyond 0.35 or so, it suggests that either the refinement is failing to deal correctly with the finer detail, or else that the intensities in this outermost part of the diffraction data are insufficiently accurate to give usable information.

Rebuilding poor parts of a model

Refinement is a cyclic process, which can be continued indefinitely.

There are almost always parts of a structure that the purely automatic procedure of refinement fails to model correctly. Sometimes, but not always, simulated annealing may

---

Box 12.5 The free *R* factor

It is now usual to compute an 'unbiased' *R* factor called R_{free}. To do this, a set of test reflections are chosen at random before refinement begins. These reflections are eliminated from the summation defined in eqn 1, Box 12.2, before any refinement is begun. There is thus a loss of experimental data for refinement and the 'safety factor' is reduced. However, the great advantage is that these observations are not available to the refinement procedure, and improvements to

$$R_{free} = \frac{\sum_{\text{test set}} \left| \left| F_{obs} \right| - \left| F_{calc} \right| \right|}{\sum_{\text{test set}} \left| F_{obs} \right|}$$

reflect genuine improvements to the model, rather than fitting to a set of data that might not be adequate to define the model.

automatically avoid the problem. At each round of refinement it is essential to review a refined model for errors. This is normally done by studying an $(F_o - F_c)$ difference map, which represents the difference between the electron density derived from the observed diffracted intensities, and that calculated from the model.

If this map has any prominent features, it gives clear evidence that a part of the model has been wrongly assigned. This part of the map needs to be re-interpreted.

Another type of map, probably less biased by incorrect features in the model, may be obtained by an 'omit' map, also explained in Chapter 11. In certain cases, reinterpretation by basing a new model on an omit map may provide a way out of a difficulty.

When is refinement complete?

If a difference map at the highest resolution shows any interpretable features, the whole refinement process is not complete.

You will probably never know how far you are from the real minimum. (You will only know this if subsequently you, or someone else, discovers that your structure is (partly) wrong.) At resolution of 2.5 Å or better, one expects to reduce R_{free} below 0.24. Even when the resolution is worse than this, if R_{free} is greater than 0.3, the results should be considered with some scepticism. If possible, modified methods of interpretation or refinement should be tried.

Example of very high resolution refinement: concanavalin A (Deacon et al. 1997)

The structure of this 25 kDa lectin, first determined in 1975, was refined at 1.6 Å resolution in 1994 for a metal-containing form. This structure (without solvent molecules) was the model for refinement of the metal-free structure to 0.94 Å resolution in 1997. The refinement is documented in detail (Fig. 12.10). In the first part of the refinement, bond lengths and angles were restrained, planar groups kept flat, and short atomic contacts prevented. Starting with 1.64 Å data, the resolution was extended cautiously. In further rounds of refinement, 320 water molecules were included and protein atoms were allowed anisotropic temperature factors (6 per atom) to achieve $R_{free} = 0.149$, $R = 0.133$. At this resolution many hydrogen atoms were clearly identifiable, but their inclusion

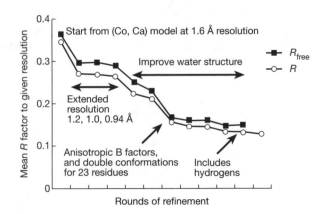

Fig. 12.10 Progress of refinement of metal-free concanavalin A at 0.94 Å resolution (data from Deacon et al. 1997).

made little difference to the final R factor ($R_{\text{free}} = 0.148$). At this stage there were still over six intensity measurements for every structural variable being refined.

In a final refinement all restraints were released. Bond lengths were thus estimated directly. This showed plainly that some restraints on bond lengths had prevented appropriate adjustment (especially in Asp and Glu side chains, where the bond length depends on carboxyl protonation). On the other hand, in more mobile parts of the structure the results deteriorated when restraints were released.

Comment

Extension of the resolution in rounds 2–4 (to include five times as many observations) took place without increasing R. This implies that all the atoms were so well placed that they remained within the sharper density distribution at the increased resolution. Changes in R during the refinement were followed closely by R_{free}, behaviour that suggests reliable changes were being made throughout. The exceptionally low R factor achieved at very high resolution not only confirms the model, but demonstrates the accuracy of the intensity measurements (already discussed in Chapter 6).

Example of difficult refinement: metabotropic glutamate receptor (Kunishima *et al.* 2000)

In this example, the authors were up against the limits of what is achievable at poor resolution. They worked on three crystal forms of the ligand-binding fragment of the receptor: two unliganded forms I and II, and a complex with glutamate. In all the crystals, the asymmetric unit contains a dimer of 2×522 amino acids, related by a non-crystallographic pseudo-2-fold axis. The unit cell of form II is similar to that of the 'complex', while form I has a completely different unit cell. Although the 'complex' diffracted to 2.2 Å resolution, intensities for the unliganded forms were measurable only to 3.7 Å or worse, and even at these resolutions the intensity measurements had become very poor. Some details of the analysis are given in Table 12.2.

Phase information was only achievable to 4.0 Å resolution. At this resolution it was confirmed that the receptor was closely similar to a known structure, the leucine/isoleucine/valine-binding protein. The main chain of this protein, appropriately placed in the 4 Å electron density for the 'complex', provided the starting-point for rigid-body refinement,

Table 12.2 Crystallographic analysis of metabotropic glutamate receptor (data from Kunishima *et al.* 2000)

	'Complex'	Free form I	Free form II
Volume of asymmetric unit (Å³)	351000	456000	352000
Resolution range (Å)	2.20	3.70	4.00
outer shell	2.25–2.20	3.83–3.70	4.14–4.00
R_{merge}	0.045	0.146	0.181
outer shell	0.137	0.520	0.250
Total number of reflections	65274	18693	10016
no. reserved to evaluate R_{free}	3497	950	534
R factor	0.196	0.244	0.254
R_{free}	0.227	0.287	0.328

which was originally restrained to satisfy the non-crystallographic symmetry. The resulting main-chain conformation was used to generate a model for the 'complex' which proved refinable at 2.2 Å resolution. This showed significant departure from the non-crystallographic symmetry and ultimately included 897 of the 1044 amino acids in the dimer.

Molecular replacement was used to indicate an appropriate orientation of the same dimeric model in unliganded form I, giving a result consistent with SIR phases from a platinum derivative of form I. This dimeric model was refined against the 3.7 Å intensity data. It was found to be more symmetric than the 'complex' form, and tight non-crystallographic symmetry restraints were applied to most of the molecule. The resulting electron-density map was ultimately refined to show the positions of 912 amino acids.

Based on this interpretation, four domains of the monomer structure were identified, such that the whole unliganded molecule could be represented as four rigid-body fragments of the 'complex' molecule, whose structure had been refined at a more satisfactory resolution.

Turning to the 4.0 Å data for unliganded form II, its structure was refined in terms of these four rigid domains. It was found to depart from dimer symmetry in a similar way to the 'complex' form. Further restrained refinements were carried out, including simulated annealing, leading to assignment of 896 amino-acid positions.

The structures of the two unliganded forms both clearly revealed the 15 α-helices observed in the complex.

Comment

The satisfactory R values at good resolution for the 'complex' form give confidence in this structure.

The rigid-body refinement of form I has few variables, and agreement with the independently determined SIR phases confirms its correctness. Details of the further refinement of form I are not given. If all bond lengths and angles are tightly restrained, the flexibility is equivalent to about 4 conformational angles for each amino acid. There would be some 3650 variables and 11 500 observed reflections (after removing some for calculation of R_{free}) giving an adequate safety factor (if one accepts the usefulness of the poorly determined reflections near the resolution limit). Refinement with tight restraints on non-crystallographic symmetry doubles this safety factor.

Refinement of form II with 896 amino acids without non-crystallographic symmetry restraints would be on the limit of uncertainty. Assuming about four variables for each amino acid of the dimer, there would have been 3500 refinable variables, with only 9500 observations to refine against. The authors were able to improve the safety factor considerably by applying non-crystallographic symmetry restraints to the inner core of the protein (N. Kunishima and K. Morikawa, personal communication). In this refinement R decreased significantly but moved away from R_{free}.

In both cases the appearance of 15 α-helices, readily identifiable at 4.0 Å, gives confidence in the general interpretation of the structure, although details of form II must be uncertain, as warned by the large ratio of R_{free} to R.

Further reading

The recommended textbook introduction to refinement is:
Stout, G.H. and Jensen, L. (1989). *X-ray structure determination. A practical guide* (2nd edn), pp. 358–78. Wiley, New York.

The free R factor is presented by

Brunger, A.T. (1992). Free R-value—a novel statistical quantity for assessing the accuracy of crystal-structures. *Nature*, **355**, 472–5.

Reviews of refinement methods:

Brünger, A.T. and Rice, L.M. (1997). Crystallographic refinement by simulated annealing. *Methods Enzymol.*, **277**, 243–69.

Jensen, L.H. (1997). Refinement and reliability of macromolecular models based on X-ray diffraction data. *Methods Enzymol.*, **277**, 353–65.

Ten Eyck, L.F. and Watenpaugh, K.D. (2001). Introduction to refinement. In *International tables for crystallography*, Vol. F. (ed. M.G. Rossmann and E. Arnold), pp. 369–74. International Union of Crystallography/ Kluwer, Dordrecht.

13

Accuracy of the final model

The result of all the work described in the previous chapters will be a set of coordinates and other data suitable for deposit in the Protein Data Bank. You or I may use these coordinates, and we need to have some insight into their accuracy and reliability. In the previous chapters, indicators have been described, which may suggest aspects of the data or interpretation procedures that might lead to problems. But as the determination of protein crystal structures becomes more routine, many of these indicators are omitted from publications.

Fortunately, crystallographic procedures are self-checking to a large extent. It is rare for a major error of interpretation to lead right through to a published refined structure. A high R_{free} factor is a warning, especially if coupled with departures from the requirements of correct bond lengths, angles, and acceptable dihedral angles.

On the other hand, there will always be a desire to squeeze more results from the data. All interpretations are subject to error; nearly all protein crystals have regions that are less ordered, where accurate interpretation is less feasible; and the structure may be over-refined, using too many variables for the data. If the majority of the molecule is correctly interpreted, a reasonable R factor may be obtained even though some small regions are completely wrong.

Independent indicators of accuracy

During refinement it is usual to restrain the bond lengths and bond angles to be near their theoretical values, as described in Chapter 12. The extent to which bond lengths and bond angles depart from these values is often quoted as an indicator of accuracy. These departures are, however, difficult to interpret because they depend on how tightly the restraints have been applied. The same applies to the restraint of certain coordinates to lie in a plane.

This difficulty illustrates a general problem. Designers of refinement procedures are understandably anxious to improve their procedures to lead directly to a well-refined structure. Every aspect of structure that can be recognized as having a regularity could, in principle, be expressed as a restraint which enforces it during refinement.

The Ramachandran plot, which shows the distribution of (ϕ, ψ) conformational angles along the polypeptide chain of a protein, can make a very good independent indicator of quality. As refinement proceeds, it is found that the distribution of conformational angles in the plot improves. Fewer conformational angles lie in the 'forbidden' regions of the plot; and the distribution within the ranges representing secondary structure becomes tighter. In a well-refined, accurate structure, the conformational angles representing α-helical structure form a very tight distribution.

Example: *Erythrina* trypsin inhibitor (Onesti 1991)

The MIR results for *Erythrina* trypsin inhibitor resulted in an electron-density map which could be interpreted confidently for 80% of the polypeptide chain (described in Chapter 7). The phase angles were improved by solvent flattening, and the structure was refined, including simulated annealing. A Ramachandran diagram derived from a model built using the MIR results is compared in Fig. 13.1 to that for the refined structure.

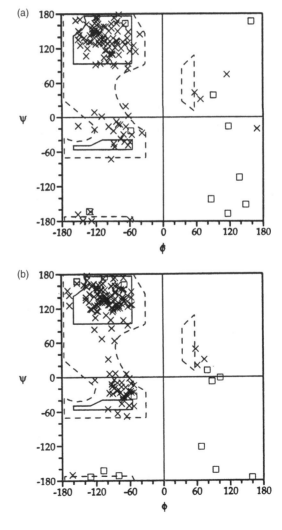

Fig. 13.1 Ramachandran diagrams (a) from a model built directly from MIR results, and (b) for the refined structure for *Erythrina* trypsin inhibitor. □—glycine; ×—other amino-acids. (Reproduced by permission from Onesti 1991.)

It may be noted that the conformational angles representing α-helical structure (those with negative φ and small negative ψ angles) form a more tightly bunched distribution in the Ramachandran diagram for the refined structure, and that several amino acids that present conformations in the 'disallowed' region outside the dashed lines in the unrefined model have been brought into 'allowed' regions. These changes are typical of the improvement made during a successful refinement.

Sooner or later, someone is bound to devise constraints that force a structure under refinement to fit the permitted regions of the Ramachandran plot. If such constraints were applied, an 'improvement' in the Ramachandran plot would cease to give independent reassurance about the refinement.

Another important indicator involves the existence of reasonable hydrogen bonds. Almost all buried polar groups are stabilized by hydrogen bonds or hydrogen-bond networks, and the geometry of these may be reviewed.

Preferred side-chain dihedral angles

Similar considerations arise from the well-known fact that side chains are often found in preferred conformations. The prevalence of favoured side-chain dihedral angles can give evidence about the quality of structure determination, provided this prevalence has not been imposed during refinement.

Disordered side chains: high *B* factors or omitted coordinates?

The conformations of long side chains which are highly exposed to solvent are expected to be very mobile. Lysine and arginine side chains are the most frequent examples. If different *B* factors are assigned to each atom, a rapid increase in *B* will be observed towards the end of many side chains. If *B* is greater than, say, 60Å^2, the atom is barely contributing to scattering beyond 3 Å resolution, and it will never be seen in any map. The estimate of its position derives from restraints rather than direct observation.

Some uncritical users of the Data Bank look only at the coordinates and ignore these high *B* factors. This can lead to misinterpretation which can be serious. In docking a ligand to a protein model, for example, it may be essential to consider the flexibility of some side-chains of the free protein.

If the *B* factor of an atom is high (above 50 or 60Å^2), its coordinates give little or no information, over and above the existence of bonds that link it to a better-ordered atom. In the effort to prevent over-interpretation of their coordinates, some early submissions to the Protein Data Bank omit these disordered atoms from coordinate lists. It can then be all too easy for the uncritical user to forget that the atoms even exist (though many model presentation programs will complain about their omission).

The Data Bank format allows an occupancy to be assigned to each atom. In well-refined structures this is particularly useful where the electron density shows that two or more different conformations exist. In this case, of course, different sets of coordinates appear for the same atoms, which may also create problems for some model presentation programs.

For atoms that are highly disordered, it is possible to enter coordinates as generated by the refinement procedure, but to set the occupancy to zero. It is nowadays more usual for each atom to be entered with the B factor assigned to it in refinement, and full occupancy. Multiple conformations will be found only in certain cases in well-refined structures. It should be expected, in all cases, that the indicated occupancies and temperature factors are those that were used in calculating structure factors to obtain the final R factor.

Main-chain disorder

Exactly the same considerations arise when the main chain is seriously disordered. It is common for several amino acids to be omitted entirely from a set of Data Bank coordinates. This happens frequently for amino acids at chain termini, for which there is often no significant density in an unbiased electron-density map. It also happens when seriously disordered loops exist within a structure.

Decisions whether or not a particular piece of density is interpretable are very subjective. And even seriously disordered side chains may be slightly easier to observe at improved resolution. As knowledge of protein structures becomes more extensive, and computer-assisted modelling more effective, confident assignments become possible based on poorer information.

Coordinate accuracy

When all these points have been considered, you are still entitled to ask how accurate the coordinates in the Data Bank may be.

In the past, the question was often addressed by use of the 'Luzzati plot' introduced by Vittorio Luzzati for evaluation of coordinates derived from fibre diffraction. Recently Durward Cruickshank has pointed out that this is a misuse of Luzzati's theoretical analysis, and has proposed a method of analysis with better foundation.

Cruickshank (1999) has extended a method he introduced in 1949, for evaluation of the accuracy of well-resolved atomic structures in small molecule crystallography, on the basis of their R factor. He has devised a 'quick and rough guide' to the precision of a coordinate derived from diffraction data. It predicts the error for an atom whose B factor is average, in the particular structure. He calls it the diffraction-component precision index, DPI. The coordinate error is proportional to the R factor (Box 13.1).

The R factor depends, of course, on the mean value of $||F_{obs}| - |F_{calc}||$. It can only be small if both the intensities have been measured accurately to give good $|F_{obs}|$, and also successful refinement has given good estimates F_{calc}.

Cruickshank has given a number of examples to show the precision of protein crystal structure analyses. Refinements using data to better than 2 Å and with R_{free} 0.2 or less may give coordinates of precision 0.2 Å or so. If the resolution is much worse than 2 Å, limitations on the refinement process mean that the precision gets much worse. At poor resolution, the R_{free} factor becomes very noticeably different from the ordinary R factor, and the estimated precision can easily deteriorate to 0.4 Å or more. This restates warnings given in Chapter 12, that at poor resolution refinement becomes unreliable because the safety factor (number of observations/number of variables being refined) becomes too low.

Box 13.1 Cruickshank's formula for the precision of a coordinate x

Cruickshank's analysis (1999) ignores any improvement to precision which may follow from constraining bond lengths and bond angles to known values, so in practice it gives a value larger than the true precision. It estimates the precision of the position of an atom which has an 'average' B factor for the particular structure. Well-ordered atoms may be placed more precisely; poorly ordered structure is likely to be interpreted worse. The 'diffraction-component precision index', σ (the standard error of position), states an error that is exceeded only one time in three. An error of 3σ should be encountered only once in 400 cases. Of the two types of formulae given by Cruickshank, only the simpler form based on R_{free} is discussed here. The formula is

$$\sigma(r, B_{\text{avg}}) = 1.73(N_i/n_{\text{obs}})^{1/2}C^{-1/3}R_{\text{free}}\, d_{\text{min}} \tag{1}$$

where N_i is the effective number of scattering atoms (including ordered solvent atoms), n_{obs} is the number of observed reflections, C is the completeness of the data. This formula shows that the coordinate error is proportional to the R factor. But to see how it responds to the resolution of the data, or the number of observations, rearrangements are needed.

Provided that all the atoms in the protein are ordered, the formula can be rearranged (Blow (2002), using the formula of Box 5.6, and assuming the mean atomic weight of a non-hydrogen protein atom is about 14.1) to:

$$\sigma(r, B_{\text{avg}}) = 0.32(1 + s)^{1/2}V_M^{-1/2}C^{-5/6}R_{\text{free}}\, d_{\text{min}}^{5/2}. \tag{2}$$

V_M is the Matthews volume (Chapter 5), and s is the fraction of refined atoms that are solvent atoms. The $(1 + s)$ term shows how the error increases as more parameters are refined. The R factor may be reduced by including more solvent atoms, but there is a compensating increase in error for more parameters. The error is almost inversely proportional to the completeness, C: if only half the available intensities are observed, the error is almost doubled. Most important of all, the error reduces rapidly if the resolution can be improved. If R_{free} can be kept the same at higher resolution, a 10% improvement in resolution will give about 30% improvement in accuracy.

Another rearrangement provides the quickest way to estimate the precision from protein crystallographic data. It depends only on the number of fully occupied atoms that represent the structure, the volume of the crystal asymmetric unit, the number of intensity observations, and R_{free}:

$$\sigma(r, B_{\text{avg}}) = 2.2N_{\text{atoms}}^{1/2}V_a^{1/3}n_{\text{obs}}^{-5/6}R_{\text{free}}. \tag{3}$$

Note that the expressions (2) and (3) only apply to proteins since they assume a mean atomic weight of about 14.1 for the scattering atoms. They include other approximations.

Using an approximate generalization of Cruickshank's analysis, Fig. 13.2 shows how the precision index varies with resolution, for a given R factor. It is perhaps surprising that the precision of the coordinates does not depend on the size of the structure. Given the

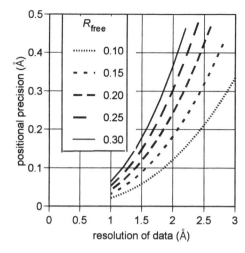

Fig. 13.2 Rough guide to precision of atomic positions using a rearrangement of Cruickshank's formula (see Box 13.1, eqn 3). This rough guide (assuming four variables per atom, 5% of complete data reserved for R_{free}, no solvent atoms added, typical V_M of 2.4 Å^3) is only indicative.

same resolution, a complete set of intensity measurements, and the same R factor, a large molecule can give coordinates just as good as a very small one.

The role of the Matthews volume, V_M, is interesting. In Chapter 5 we saw that this reflects the solvent content of the crystals. A large solvent content is generally associated with greater disorder in the crystals, limiting the available resolution and increasing the individual B factors. These factors make it difficult to achieve a high-resolution structure determination. But since it also increases the size of the unit cell, the number of orders of diffraction observed at given resolution are increased, and this leads to an improvement in the presumed precision at some particular resolution.

Another estimate of the precision of protein coordinates is sometimes available, which at least places an upper limit on errors. This is to compare two completely independent structure determinations of the same protein molecule. Such comparisons indicate precisions at least as good as those given by the Cruickshank formula.

Example: independent determinations of the same crystal structure (Daopin *et al.* 1994)

Unknown to each other, two groups independently refined the structures of similar crystal forms of transforming growth factor-β2, a 25 kDa dimeric cytokine. Comparing the refined coordinates, C$^\alpha$ positions agree to 0.10 Å, and the average difference for the best ordered 90% of atoms is 0.33 Å. There is a strong correlation between the size of coordinate differences and the B factor of the corresponding atom. Cruickshank's formula yields errors of 0.16 Å and 0.24 Å in the two structures, for each coordinate of an atom with average B factor. Theoretically this would lead to an average coordinate difference of 0.29 Å for an atom with average degree of order, agreeing well with the observed 0.33 Å.

When structure determinations of the same molecule are made in different crystal forms, it can be argued that the differences are partly due to different conformations in the different environments of the unrelated crystal forms, but this often seems to be a small effect, less than 0.1 Å on average for well-ordered atoms.

Table 13.1 Analysis of crambin at 0.54 Å resolution

Number of independent reflections[a]	112293
Number of reflections with $I > 3\sigma$[a]	79868
R_{merge}[a]	0.055
Free R factor[a]	0.096
Median atomic B factor[a]	2.5 Å^2

Refinement with spherical atoms, free bond lengths and angles

non-H atoms (anisotropic B factors)	327
H atoms (isotropic B factors)	504
Approximate number of parameters	5300
Safety factor	21

[a] Data from Jelsch *et al.* (2000).

Example, the best resolution so far: crambin at 0.54 Å (Jelsch *et al.* 2000)

The highest-resolution structure determination for a macromolecule recorded at present is that of the 46 amino-acid plant toxin, crambin, at 0.54 Å. Its crystals have a very low average B factor, of about 2.5 Å^2. Refinement without restraint of bond lengths and angles, using anisotropic temperature factors for all non-hydrogen atoms and free coordinates for each hydrogen atom, still gives a safety factor of 21 (Table 13.1). This allows ample freedom for more sophisticated analysis in terms of bonding orbitals and study of distortions of electron distribution. Bond lengths and bond angles are now accurately accessible from protein structure analysis (previously they were taken from small-molecule structures), and regular variations can be detected depending on ionization or polarization.

Further reading

An extensive review of some of the issues discussed in this chapter (and in other chapters) is given by
Kleywegt, G.J. (2000). Validation of protein crystal structures. *Acta Cryst.*, **D56**, 249–65.

The Protein Data Bank is described in
Berman, H.M. *et al.* (2000). The Protein Data Bank. *Nucleic Acids Res.*, **28**, 235–42.

A brief account of the Ramachandran plot may be found in
Stryer, L. (1995). *Biochemistry* (4th edn), pp. 420–1. Freeman, New York.

A preliminary and simplified account of Cruickshank's recent work:
Cruickshank, D.W.J. (1996). Macromolecular refinement. *Proceedings of the CCP4 study weekend* (ed. E. Dodson, M. Moore, A. Ralph, and S. Bailey), pp. 11–22. Daresbury Laboratory, Warrington.

A clearly presented full account:
Cruickshank, D.W.J. (2001). Coordinate uncertainty. In *International tables for crystallography*, Vol. F. (ed. M.G. Rossmann and E. Arnold), pp. 403–14. International Union of Crystallography/Kluwer, Dordrecht.

References

Ban, N., Nissen, P., Hansen, J., Capel, M., Moore, P.B., and Steitz, T.A. (1999). Placement of protein and RNA structures into a 5 Å-resolution map of the 50S ribosomal subunit. *Nature*, **400**, 841–7.

Ban, N., Nissen, P., Hansen, J., Moore, P.B., and Steitz, T.A. (2000). The complete atomic structure of the large ribosomal subunit at 2.4 Å resolution. *Science*, **289**, 905–20.

Bijvoet, J.M. (1949). Phase determination in direct Fourier synthesis of crystal structures. *Kon. Ned. Akad. Wetenschap.*, **52**, 313–14.

Bijvoet, J.M., Peederman, A.F., and van Bommel, J.A. (1951). Determination of the absolute configuration of optically active compounds by means of X-rays. *Nature*, **168**, 271–2.

Blow, D.M. (2002). A rearrangement of Cruickshank's formulae for the diffraction-component precision index. In press, *Acta Cryst.*, **D**.

Bourne, P.C., Isupov, M.N., and Littlechild, J.A. (2000). The atomic-resolution structure of a novel bacterial esterase. *Structure*, **8**, 143–51.

Bragg, Sir L. (1943). *The history of X-ray analysis*. British Council/Longmans Green, London.

Bragg, W.L. (1913). The structure of crystals as indicated by their diffraction of X-rays. *Proc. Roy. Soc.*, **A89**, 248–77.

Britton, K.L. *et al.* (2000). The crystal structure and active site location of isocitrate lyase from the fungus *Aspergillus nidulans*. *Structure*, **8**, 349–62.

Caspar, D.L.D. (1980). Movement and self-control in protein assemblies. Quasi-equivalence revisited. *Biophys. J.*, **32**, 103–38.

Chen, L. and Sigler, P.B. (1999). The crystal structure of a GroEL/peptide complex: Plasticity as a basis for substrate diversity. *Cell*, **99**, 1–20.

Clemons, W.M. Jr, May, J.L.C., Wimberly, B.T., McCutcheon, J.P., Capel, S., and Ramakrishnan, V. (1999). Structure of a bacterial 30S ribosomal subunit at 3.5 Å resolution. *Nature*, **400**, 833–40.

Cruickshank, D.W.J. (1999). Remarks about protein structure precision. *Acta Cryst.*, **D55**, 583–601.

Cullis, A.F., Muirhead, H., Perutz, M.F., Rossmann, A.G., and North, A.C.T. (1961). The structure of haemoglobin. VIII. A three-dimensional Fourier synthesis at 5.5 Å resolution: determination of the phase angles. *Proc. Roy. Soc.*, **A265**, 15–38.

Daniels, D.L., Cohen, A.R., Anderson, J.M., and Brünger, A.T. (1998). Crystal structure of the hCASK PDZ domain reveals the structural basis of class II PDZ domain target recognition. *Nat. Struct. Biol.*, **5**, 317–24.

Daopin, S., Davies, D.R., Schlunegger, M.P., and Grütter, M.G. (1994). Comparison of two crystal structures of TGF-β2: the accuracy of refined protein structures. *Acta Cryst.*, **D50**, 85–94.

Deacon, A. *et al.* (1997). The structure of concanavalin A and its bound solvent determined with small-molecule accuracy at 0.94 Å resolution. *J. Chem. Soc. Farad. Trans.*, **93**, 4305–12.

Dickinson, R.G. and Raymond, A.L. (1923). The crystal structure of hexamethylene tetramine. *J. Am. Chem. Soc.*, **45**, 22–9.

Diederichs, K. and Karplus, P.A. (1997). Improved *R*-factors for diffraction data analysis in macromolecular crystallography. *Nature Struc. Biol.*, **4**, 269–75.

Doublié, S., Xiang, S., Gilmore, C.J., Bricogne, G., and Carter, C.W. Jr (1994). Overcoming non-isomorphism by phase permutation and likelihood scoring: solution of the TrpRS crystal structure. *Acta Cryst.*, **A50**, 164–82.

Esnouf, R.M. (1997). An extensively modified version of Molscript that includes greatly enhanced coloring capabilities. *J. Mol. Graphics*, **15**, 133–8.

Ferré-D'Amaré, A.R., Zhou, K., and Doudna, J.A. (1998). Crystal structure of a hepatitis delta virus ribozyme. *Nature*, **395**, 567–74.

Garman, E. and Schneider, T.R. (1997). Macromolecular crystallography. *J. Appl. Cryst.*, **30**, 211–37.

Goldberg, J.D. (1993). The crystal structure of D-glycerate dehydrogenase. Doctoral thesis, Imperial College, London, UK.

Green, D.W., Ingram, V.M., and Perutz, M.F. (1954). The structure of haemoglobin. IV. Sign determination by the isomorphous replacement method. *Proc. Roy. Soc.*, **A225**, 287–307.

Hammond, C. (1997). *The basics of crystallography and diffraction*. International Union of Crystallography/ Oxford University Press, Oxford.

Harburn, G., Taylor, C.A., and Wellberry, T.R. (1975). *Atlas of optical transforms*. Bell, London.

Harrison, S.C. (1980). Protein interfaces and intersubunit bonding. *Biophys. J.*, **32**, 139–53.

Harrison, S.C., Olson, A.J., Schutt, C.E., and Winkler, F.K. (1978). Tomato bushy stunt virus at 2.9 Å resolution. *Nature*, **276**, 368–73.

Haüy, R.J. (1801). *Traité de minéralogie*, Vol. V, Pl. III, Fig. 17. Paris.

Hendrickson, W.A. and Teeter, M.M. (1981). Structure of the hydrophobic protein crambin determined directly from the anomalous scattering of sulfur. *Nature*, **290**, 107–13.

Hendrickson, W.A., Horton, J.R., and Le Master, D.M. (1990). Selenomethionyl proteins produced for analysis by multiwavelength anomalous scattering (MAD): a vehicle for direct determination of three-dimensional structure. *EMBO J.*, **9**, 1665–72.

Holmes, K.C. and Blow, D.M. (1966). The use of x-ray diffraction in the study of protein and nucleic acid structure. In *Methods Biochem. Anal.*, **13**, 113–239.

Ishikawa, K., Nakagawa, A., Tanaka, I., Suzuki, M., and Nishihira, J. (2000). The structure of human MRP8, a member of the S100 calcium-binding protein family, by MAD phasing at 1.9 Å resolution. *Acta Cryst.*, **D56**, 559–66.

Ito, N. (1991). The structure determination of galactose oxidase by multiple isomorphous replacement with anomalous scattering. In *Isomorphous replacement and anomalous scattering* (ed. W.Wolf, P.R. Evans, and A.G.W. Leslie). Proceedings of a CCP4 study weekend, pp. 116–24. Daresbury Laboratory, UK.

Ito, N. *et al.* (1991). Novel thioether bond revealed by a 1.7 Å crystal structure of galactose oxidase. *Nature*, **350**, 87–90.

Jelsch, C., Teeter, M.M., Lamzin, V., Pichon-Pesma, V., Blessing, R.H., and Lecomte, C. (2000). Accurate protein crystallography at ultra-high resolution: valence electron distribution in crambin. *Proc. Natl Acad. Sci.*, *USA*, **97**, 3171–6.

Johnston, I. (1989). *Measured tones. The interplay of physics and music.* Hilger, Bristol.

Kekulé von Stradonitz, F.A. (1866). *Lehrbuch der organischen Chemie.* Enke, Erlangen.

Kunishima, N. *et al.* (2000). Structural basis of glutamate recognition by a dimeric metabotropic glutamate receptor. *Nature*, **407**, 971–7.

Kwong, P.D., Wyatt, R., Robinson, J., Sweet, R.W., Sodoroski, J., and Hendrickson, W.A. (1998). Structure of an HIV gp120 envelope glycoprotein in complex with the CD4 receptor and a neutralizing human antibody. *Nature*, **393**, 648–59.

Mancia, P. and Evans, P.R. (1998). Conformational changes on substrate binding to methylmalonyl CoA mutase and new insights into the free radical mechanism. *Structure*, **6**, 711–20.

Matthews, B.W. (1968). Solvent content of protein crystals. *J. Mol. Biol.*, **33**, 491–7.

Merritt, E.A. and Bacon, D.J. (1997). Raster3D: Photorealistic molecular graphics. *Methods Enzymol.*, **227**, 505–24.

Olson, W.K. and Flory, P.J. (1972). Spatial configurations of polynucleotide chains. I. Steric interactions in polyribonucleotides: a virtual bond model. *Biopolymers*, **11**, 1–23.

Onesti, S. (1991). Crystal structure of *Erythrina* trypsin inhibitor. PhD thesis, Imperial College, London.

Onesti, S., Brick, P., and Blow, D.M. (1991). Crystal structure of a Kunitz-type trypsin inhibitor from *Erythrina caffra* seeds. *J. Mol. Biol.*, **217**, 153–76.

Pallardy, G., Pallardy, M.J., and Wackenheim, A. (1989). *Histoire illustrée de la radiologie.* Roger Dacosta, Paris.

Perutz, M.F. (1992). *Protein structure: new approaches to disease and therapy.* Freeman, New York.

Ramachandran, G.N., Ramakrishnan, C., and Sasisekharan, V. (1963). Stereochemistry of polypeptide chain conformations. *J. Mol. Biol.*, **7**, 95–9.

Rossmann, M.G. and Blow, D.M. (1962). The detection of sub-units within the crystallographic asymmetric unit. *Acta Cryst.*, **15**, 24–31.

Rossmann, M.G. *et al.* (1985). Structure of a human common cold virus and functional relationship to other picornaviruses. *Nature*, **317**, 145–53.

Saenger, W. (1984). *Principles of nucleic acid structure.* Springer-Verlag, Berlin.

Sigler, P.B., Jeffery, B.A., Matthews, B.W., and Blow, D.M. (1966). An x-ray diffraction study of inhibited derivatives of chymotrypsin. *J. Mol. Biol.*, **15**, 175–92.

Simpson, A.A., Chipman, P.R., Baker, T.S., Tijssen, P., and Rossmann, M.G. (1998). The structure of an insect parvovirus (*Galleria mellonella* densovirus) at 3.7 Å resolution. *Structure*, **6**, 1355–67.

Taylor, C.A. and Lipson, H. (1964). *Optical transforms*. Bell, London.

Teeter, M.M. (1999). Disorder patterns and peptide geometry in crambin at 0.54 Å resolution. *Biophys. J.*, **76**, A387.

Teplyakov, A., Obmolova, G., Badet-Denisot, M.A., Badet, B., and Polikarpov, I. (1998). Involvement of the C terminus in intramolecular nitrogen channeling in glucosamine-6-phosphate synthase: evidence from a 1.6 Å crystal structure of the isomerase domain. *Structure*, **6**,1047–55.

Tocilj, A. *et al.* (1999). The small ribosomal subunit from *Thermus thermophilus* at 4.5 Å resolution: pattern fittings and the identification of a functional site. *Proc. Natl Acad. Sci. USA*, **96**, 14252–7.

van't Hoff, J.H. (1874). *Structuur-formules in de Ruimke*. Greven, Utrecht.

Vrielink, A. (1989). The crystal structure determination of cholesterol oxidase. Doctoral thesis, Imperial College, London, UK.

Watenpaugh, K.D., Sieker, L.C., Herriott, J.R., and Jensen, L.H. (1973). Refinement of the model of a protein. Rubredoxin at 1.5 Å resolution. *Acta Cryst.*, **B29**, 943–56.

White, A., Ding, X.C., van der Spek, J.C., Murphy, J.R., and Ringe, D. (1998). Structure of the metal-ion-activated diphtheria toxin repressor tox operator complex. *Nature*, **394**, 502–6.

Wilson, A.J.C. (1942). Determination of absolute from relative x-ray intensity data. *Nature*, **150**, 151–2.

Wimberly, B.T. *et al.* (2000). Structure of the 30S ribosomal subunit. *Nature*, **407**, 327–39.

Woolfson, M.M. and Fan, H.F. (1995). *Physical and non-physical methods of solving crystal structures*. Cambridge University Press, Cambridge.

Yu, R.C., Hanson, P.I., Jahn, R., and Brünger, A.T. (1998). Structure of the ATP-dependent oligomerization domain of *N*-ethylmaleimide sensitive factor complexed with ATP. *Nat. Struct. Biol.*, **5**, 803–10.

Zhang, K.Y.J. and Main, P. (1990). The use of the Sayre equation with solvent flattening and histogram matching for phase extension and refinement of protein structures. *Acta Cryst.*, **A46**, 41–6.

Index

| | *see* modulus, complex number

absorption edge 147–9, 153, 157–60
  K edge 147–8, 153
  L edge 147–8, 157
accuracy, final model 225–8
amplitude, structure, *see* structure
    amplitude
  wave 5, 45–6, 52
anomalous scattering 147–61
  see also isomorphous replacement
Argand diagram 48–50, 126, 131,
    142
Arndt, Uli 94
asymmetric unit 35–8, 41, 100, 180
atomic scattering factor 92, 148, 150

B-factor 91–2, 106, 112–14, 117–18, 199,
    213–14, 228
  anisotropic 213–14, 228
beam stop 19, 57, 92
Bernal, Desmond vii, 94
best phase, best structure factor 139,
    142–4, 191
Bijvoet, Johannes 147, 150–1
Bijvoet difference 151, 154, 157–9
body-centred unit cell 35, 187
Booth, Alan 208
boundaries, molecular 162, 186
Bragg, Lawrence 8, 10, 21, 70, 82–4
Bragg, William 10
Bragg's law 83–4, 89–90

centrosymmetry 24, 62–5, 96
  projections 122–6
characteristic radiation 12
charge-coupled device (CCD) 18–19,
    106, 117–18
coherent scattering 15, 147
collimator 19
completeness, of intensity measurement
    111, 115–19, 167, 226

complex conjugate 49–50, 64, 150–3
complex number 47–51, 148–9
conformational angles 197, 210–11, 213
contour maps 16, 192–4
conventions
  crystal axes 32–41
  Fischer 9, 151
  unit cells 35–6, 39–40, 99
convolution 72–80, 128, 165
coordinates 35–6, 47, 61, 80, 91, 164,
    171, 205, 222–225
  omitted 224
correlation coefficient 167, 188–9
cross vector 165–7
Cruickshank, Durward 225–7
crystal
  axes, conventions 32–41
  growth 30, 128
  imperfections 30, 77
  lattice 29–35, 72–6, 80, 87–8
  size 15
  systems 33–4
crystallographic point group 25, 27
cubic symmetry 25, 28, 83, 186–7

data collection strategy 105, 108–10
  American method 108
  Laue method 96–7, 109
de Broglie relation 20
Debye, Peter 91
Debye-Waller factor, *see B*-factor
density modification 177–190
detector, X-ray 16–9, 62, 106
difference map
  anomalous, Patterson 158
  electron density 134–6, 162, 175
  Patterson 130–4
diffraction grating 58–60, 67
diffraction-component precision index
    225–7
diffractometer 17, 115
dihedral angles 210–11

disc, diffraction by 70–1
disorder 77, 91–2, 105–6, 112–17, 142,
    195, 198, 209, 224–6
double-spike function 69–72

electromagnetic waves 4–5, 10–11, 45,
    62, 122
electron 12–3, 19–20, 57, 147–8
electron density maps 142–6, 174–5, 178,
    185–6, 189–204
  difference map 134–5, 161, 175, 190,
      198–203, 218
  $F_o - F_c$ map 175, 199–203, 218
  $2F_o - F_c$ map 200–3
  omit map 201–3, 218
  resolution and 195–7
electron microscopy 15–16, 20, 29–30,
    190
equivalent positions 38–9, 42–3
Escher, Maurits 44
Euler, Leonhardt 33, 169
Eulerian angles 169–70
Ewald, Paul 82
EXAFS 148
exp, notation 50

face-centred unit cell 35–6, 39–40
FAST detector 17–18
Fedorov, Evgraph von 41
figure of merit 139–46, 156, 191, 194
Fischer, Emil 8, 9, 151
Fourier, Joseph 51, 57–8
Fourier coefficients 52, 55–6, 64–5, 73,
    132, 199–200
Fourier series 52, 55–6, 66, 73, 134
Fourier transform 54, 56, 61–81, 91–2,
    122, 192
Fraunhofer diffraction 58, 68
frequency, wave 47
Friedel's law 63–4, 94, 97–8, 147, 150–3
Friedrich, Walter 8–9, 95

gamma-rays 5
Gaussian function
  Fourier transform 68–70, 76–7, 80,
      91–2
  representing crystal disorder 76, 112
  representing experimental error 140
generator, X-ray 12–4, 106

handedness 136
Harker diagram 137–40
harmonic 51, 53, 94

hexagonal symmetry 33
histogram 178–9
histogram matching 178–80, 202–3
Hodgkin, Dorothy vii
hydrogen bonds 4, 224

i operator 48
$<I>/<\sigma(I)>$ statistic 113, 117–19, 159–60
image plate 17–18, 106
incoherent scattering 15, 109
independent reflections, see unique
    structure factors
intensity
  data quality 110, 115–19
  scattered 105–6, 110–11, 112–17
  wave 16, 45–51
intermolecular vectors 166–7
intramolecular vectors 166–7
inversion symmetry 24–5
ionizing radiation 12
isomorphism 124, 142
isomorphous derivative 121, 123–9,
    144–6, 187
isomorphous replacement 121–46, 186–7
  equation 125
  multiple (MIR) 128, 135–44, 170
  phase determination by 127–8, 137–46
  sign determination by 125–6
  single (SIR) 127, 154, 220
  with anomalous scattering, see SIRAS
      and MIRAS

Kekulé von Stradonitz, August 8–9
Knipping, Paul 8–9, 95

lack of closure 140–3
lattice plane 85–90, 94–7
lattice symmetry 30–3, 35, 37, 41, 84, 97
lattice translation 32–9, 42, 132, 165, 184
Laue, Max von 8, 10, 82, 95
Laue diffraction 58, 96–7, 109
Le Bel, Joseph 8
least squares, method of 206, 209
Leeuwenhoek, Antonie von 4
Lipson, Henry 70
local symmetry 181, 185, 220
Luzzati, Vittorio 225

MAD (multiple wavelength anomalous
    dispersion or diffraction) 157–60
magnitude, complex number 47, 51, 126
maps, see electron density; Patterson
Matthews volume 100, 162–3, 226–7

maximum entropy 190
microbeam X-ray source 13–14
MIR (multiple isomorphous replacement) 128, 136–146, 170
MIRAS (multiple isomorphous replacement with anomalous scattering) 155–8, 178, 210, 216
mirror, reflection of waves 83–4
mirror symmetry 23–4
model bias 201
model building, modelling 10, 197–200, 211, 217, 225
model, mathematical 206
model structure 162–75, 182, 210–12, 216
  accuracy 222–8
modulus, complex number 47–8, 126
molecular averaging 185
molecular replacement 162–76, 180–1, 216, 219–20
monochromatic X-rays 58
monochromator 18–19
monoclinic symmetry 33–4, 38–40, 43–4
moving-crystal method 93
multi-wire detector 18
multiple conformations 224–5
multiplicity of intensity measurement 115–17
multiple isomorphous replacement, see MIR, MIRAS
multiple wavelength anomalous diffraction or dispersion 157–60

neutron 14, 19–20
  detectors 62
Newton, Isaac 10
non-crystallographic symmetry 181, 185, 220
non-isomorphism 124

observational equation 212
occupancy 224
oligomers, symmetry 25–6, 180–1, 185–6
omit map 201–3, 218
omitted coordinates 224–5
operation, see rotation, symmetry, translation
operator i 48
optical diffraction 70–1, 75–7
origin, choice of 37–9, 41–3, 46, 131–4, 164, 170–1
orthogonality rule 54–5, 73–4
orthorhombic symmetry 33–4, 40–2

Patterson, Lyndo 77
Patterson function 77–80, 164–73, 181–5
Patterson maps
  anomalous difference 158
  difference 130–4, 144
Perutz, Max vii, 3, 121
phase angles 46, 60–2, 66, 95, 127
  calculated 91, 133–6, 173–5, 189, 199–200, 214
  combined 178, 190
phase, best 139, 142–4, 191
phase error 139, 142, 144, 155–6, 187, 194
phase extension 188–9
phasing power 139, 142–6, 155–6, 160
photographic film 16–17
photon 10–1, 15–16, 18–19
Planck's constant 11, 20
Plato 28
point-group symmetry 25–8
position-sensitive detector 17
primitive unit cell 33, 35, 37, 40, 97–9
proportional counter 17
Protein Data Bank 222–5
pseudo-symmetry 28

quantum theory 10–11, 19

R-factor 110–11, 188–9, 208, 215–20, 225
  R_{anom} 157–9
  R_{deriv} 128, 156–8, 187
  R_{diff} 128–9
  R_{free} 217–22, 225–8
  R_{iso} 128–9
  R_λ 158–9
  R_{meas} 110
  R_{merge}, R_{sym} 110–11, 128–9
radiation damage 12, 15–16, 20, 105–7
Ramachandran angles 210–11, 213
reciprocal lattice 75
refinement 205–21
  classical method 208
  convergence 216
  general principles 207
  refinement parameter 205
  rigid body 216, 219–20
reflection symmetry 23–4
reliability factor, see R-factor
resolution 4, 19, 63, 67, 76–7, 89, 93, 98, 107, 115, 228
  appearance of electron density 195–7
  in refinement 209, 214–15, 219–21, 228

resolution limit 77, 101, 109, 112, 113, 181
restraints in refinement 210–16, 219–20, 222, 224
rhombohedral symmetry 33(footnote)
right-handed coordinate system 35
rigid displacement 163–4, 216
Röntgen, Wilhelm 16
Rossmann, Michael 108
Rossmann fold 16
rotating crystal measurement 93–4
rotating-anode X-ray generator 13, 18, 19
rotation function 165–70
    self-rotation function 181–5
rotation operation 21–4, 163–9, 173–4, 181
    Eulerian system 169–70, 184
    κ, ψ, φ system 163–4, 169, 181–5
rotational symmetry 22–5, 98–9

safety, X-ray 12
safety factor, in refinement 209, 213–16, 220, 225, 228
Schoenflies, Arthur 41
screw axis 39–41, 43–4, 98–9, 122
selenium
    absorption spectrum 148
    anomalous scattering spectrum 153
    in selenomethionine 158–9
self vectors 166–7
self-rotation function 182–5
self-translation function 185
sigma weighting 200, 204
simulated annealing 216
sinusoidal wave 5, 45–6, 51
SIR (single isomorphous replacement) 127, 154, 220
SIRAS (single isomorphous replacement with anomalous scattering) 154, 158
solid state detector 18–19
solvent
    in crystals 30
    inclusion in models 201, 209, 215, 218
solvent flattening 178, 180, 186
space group 35, 37–42
stereoscopic viewing 16, 193
structural refinement, see refinement
structure amplitude 53
structure factor 52, 61, 64, 74, 91, 98
subunit 173, 180–9
symmetry 22–44
    inversion 24–5
    lattice, see crystal lattice
    local 181, 185

mirror 23–4
    non-crystallographic 181, 185
    of oligomers, see oligomers
    point group 25–8
    reflection 23–4
    rotational 22–5, 98–9
    screw 39–41, 43–4
    space group 35, 37–43
synchrotron 13–14, 96, 105, 116–19
systematic absences 98–9

Taylor, Charles 70
temperature factor, see B-factor
temperature, of crystal 107–8
thermal vibration 76, 92, 114
top-hat function 68–70
torsion angles 197, 210–11
transient states 97, 109
translation function 170–3
    self-translation function 185
translation operation 31–3, 163–4, 173
triclinic symmetry 33–4, 37
trigonal symmetry 33

unique structure factors 99, 111
    number of 99, 101, 115–19, 226, 228
unit cell 32–3, 35–43, 90, 100
    conventions 35–6, 39–40, 99
    see also body-centred, face-centred, primitive

van't Hoff, Jacobus Henricus 8–9
velocity, wave 11, 46–7
viruses, spherical 28, 30, 181, 186–9

Waller, Ivar 92
wavelength 4–6, 13–16, 19–20, 45–7
Wellberry, Thomas 70
white radiation 12, 95–7
Wilson, Arthur 114
Wilson temperature factor plot 113–14
Wonacott, Alan 94

X-ray detector 16–19, 62, 106
X-ray generator 12–14, 106
X-ray microscope 7
X-rays, monochromatic 58
X-rays, white 12, 95–7

ytterbium, anomalous scattering spectrum 157

zone plate 7